Lab Manual to Accompany

THE SCIENCE OF AGRICULTURE
A BIOLOGICAL APPROACH

4TH EDITION

RAY V. HERREN
CATHERINE A. TEARE KETTER

Lab Manual to Accompany

The Science of Agriculture: A Biological Approach

Fourth Edition

Ray V. Herren

Catherine A. Teare Ketter

DELMAR
CENGAGE Learning™

Australia • Canada • Mexico • Singapore • Spain • United Kingdom • United States

DELMAR
CENGAGE Learning™

Lab Manual to Accompany
The Science of Agriculture: A Biological Approach, Fourth Edition
Ray V. Herren and Catherine A. Teare Ketter

Vice President, Editorial: Dave Garza

Director of Learning Solutions: Matt Kane

Senior Acquisitions Editor: Sherry Dickinson

Managing Editor: Marah Bellegarde

Senior Product Manager: Christina Gifford

Editorial Assistant: Scott Royael

Vice President, Marketing: Jennifer Baker

Marketing Director: Debbie Yarnell

Marketing Manager: Erin Brennan

Marketing Coordinator: Erin DeAngelo

Senior Production Director: Wendy Troeger

Production Manager: Andrew Crouth

Senior Content Project Manager: Katie Wachtl

Senior Art Director: Dave Arsenault

Library of Congress Control Number: 2010941450

ISBN-13: 978-1-4390-5774-2

ISBN-10: 1-4390-5774-5

Delmar
5 Maxwell Drive
Clifton Park, NY 12065-2919
USA

NOTICE TO THE READER

Printed in the United States of America
1 2 3 4 5 15 14 13 12 11

FD087

CONTENTS

INTRODUCTION

Notes to the Student

The laboratory exercises contained in this manual are not designed to be completed in one class period. Where possible, laboratory methods used by agricultural scientists and biological scientists are used to give you an authentic feel for the kinds of tests these professionals do. The scientific method is a model for problem solving used by scientists to generate solutions to real-world problems. Many of the lab exercises will require you to use the scientific method including recording data on a daily or repeated basis, analyzing results, and making recommendations about practical solutions to the problems posed. A few of the exercises in this manual are appropriate for an Advanced Placement or College Preparatory Program. Many of the experiments could be expanded for science fair projects, 4-H, FFA, or agricultural competitions. It is our hope that you enjoy applying the knowledge gained in the Agricultural Science classroom to laboratory protocols and problem solving.

A Word about Lab Safety

Chemicals and equipment are used in almost all laboratory settings. With proper handling and technique, the equipment, organisms, and chemicals used in these lab exercises are safe. It is important that you follow all of the directions carefully. Pay particular attention to the "Safety Notes" where they appear. The "Safety Notes" contain special precautions for handling chemicals and living organisms. Do not eat or drink in the lab classroom and always wash your hands at the beginning and end of each class period. Wear safety goggles and use gloves when appropriate. Students with hair below the ears should secure their hair before working with any open flame (like an alcohol lamp or bunsen burner). While completing the lab activities, students should wear long pants that completely cover the legs and leather shoes that enclose the feet. Some of the lab exercises will require field trips for the collection of samples. Dress suitably for any field work and follow all of the precautions outlined by your instructor. Remember that science is fun, but for everyone to be safe, all of the rules of conduct in the classroom must be followed.

New to this Edition

Three new labs have been created for this latest edition to correspond with new content in The Science of Agriculture, 4th edition. The labs are Water Safety, Wildlife Habitat Suitability, and Energy from Agriculture.

ACKNOWLEDGMENTS

The following individuals provided invaluable assistance and deserve a special word of thanks. Dr. Denise I. Bounous, University of Georgia College of Veterinary Medicine-Pathology provided insights into large animal renal function and the pathophysiology of renal disease. The late Dr. Walstine L. Steffens, University of Georgia College of Veterinary Medicine-Pathology supplied the original electron micrographs used in this manual. Dr. David C. Coleman, Institute of Ecology University of Georgia, contributed reference materials and discussion on soil ecology and soil methodology. David Hart, formerly School of Forest Resources, University of Georgia, provided copies of Boyd's work on water quality in fisheries production. Margi Flood, Gainesville College and State University, Gainesville, Georgia, loaned Wetzel and Wetzel and Liker's limnological works to assist in the research for Chapter 17. Dr. Russell Malmberg, Plant Biology Department, University of Georgia, for his assistance with Exercise 9; Dr. Nick Fuhrman for his creation of Exercise 19.

I would like to acknowledge all of the encouragement I have received from Dr. William H. Darden, Jr., Dr. Thomas Baumar, Dr. Gale Davis, and Dr. Joseph Scheiring, Biology Department, University of Alabama, for 14 years of guidance, learning, and enthusiasm during my tenure at Alabama. It is my hope that the exercises in this volume will inspire students to study agriscience with the same appetite and enthusiasm.

DEDICATION

This book is lovingly dedicated to my husband, Dennis, and my son, Andrew, who spent many hours in one another's company while I researched and wrote this book. Thank you for your understanding and patience.

—Catherine A. Teare Ketter

EQUIPMENT LIST

The following equipment supplies were carefully selected to be used with the exercises in the Lab Manual to Accompany the Science of Agriculture: A Biological Approach, 4th edition. All items listed here are available from Carolina Biological Supply Company, 2700 York Road, Burlington, NC 27215. You may also place your order or request a catalog by calling toll free 1-800-334-5551. Obviously, there are alternatives to many of these items. This list was provided for your convenience only. It is not intended as an order form.

		Carolina Biological Supply Catalog Number
Exercise 1		
Calculator		91-2442
Metric Ruler 6″		
Metric Ruler 12″		70-2612
Meter Stick		70-2624
Metric Balance		70-2153
Erlenmeyer Flask	(500 ml)	72-6514
	(1,000 ml)	
Florence Flask		72-6056
Beaker (set) w/	(50 ml)	961963
	(100 ml)	961963
	(600 ml)	961963
Graduated Cylinder (set)	(10 ml)	961865
	(50 ml)	961865
	(100 ml)	961865
	(1,000 ml)	961865
Thermometer		Multiple available
Exercise 2		
Aquarium		67-0152
32 grams of Algae		15-1710

Editor's Note: Since 32 grams can be costly, an alternative suggestion is to grow your own algae.

Exercise 3		
Calculator		91-2442
Compound Microscope		
	(Monocular)	59-1262
	(Binocular)	59-1004

	Carolina Biological Supply Catalog Number
Dissecting Microscope	59-1822
Letter "E"	(Various available)
Threads	(various available)
Clear Plastic Metric Ruler	70-2603
Slides	63-2950
Cover Slips	63-2960
Lens Paper	63-4005
Pond Water	163380

Exercise 4

Gravel (and sand)	163000
Sand (and gravel)	163000
Loam	
Silt	
Clay	
pH Electrode	185748
pH Paper	895511
Soil Sample 1	
Soil Sample 2	
Soil Sample 3	
Sieves	GE09301
Bag	
Canvas Bag	

Exercise 5

Compound Light Microscope	59-1000
Rhizobium (N2 Fixing Bacteria)	155270
Slides:	
Mixed Blue-green Algae	294906
Mixed Diatoms	151287
Mixed Bacterial Types	(various available)
Soil Type A	
Soil Type B	
Soil Type C	
Sterile Pipette	215060
Slides	63-2960
Parafilm to seal Petri Dish	215600
Dissecting Needle	627220
Immersion Oil	85-2980
Lens Paper	63-4005
Crystal Violet Stain	892261
Sterile Flask w/10 ml Sterial H20	19-9982

	Carolina Biological Supply Catalog Number
16 × 150 ml S/C Test Tube w/9 ml Sterial H2O	
Pipette	
Sterile Petri Dish	
Nutrient Agar Bottle	77-6360
Dissecting Microscope	59-1822
Magnifying Lens	60-2221
Alcohol	86-1261
Culture Dish	74-1004
Safety Pipette Filler	

Exercise 6

Azolla	16-1800
Gram Stain Kit	
Prepared Slide/Mixed Bacterial/ Gram Stain	
Oil Immersion	85-2980
Prepared Slide w/Endospores	
Culture Dish	74-1004
Nutrient Agar Plates	82-1862
Grease Pencil	65-7730
Ampicillin Disks	
Penicillin Disks	
Streptomycin Disks	80-6376
Tetracycline Disks	80-6476
E. Coli Culture	15-4921
B. Subtilis Culture	15-4921
Metric Ruler	70-2605
Mortar and Pestle	74-2904
Azolla	16-1800
Slides	63-2950
Cover Slips	63-2960
Cyanobacteria Mixture	19-9980

Exercise 7

Slides	
Cover Slips	63-2960
Amoeba	131085
Volvox	15-2665
Depression Slide	63-2200
Yeast Suspension	15-6251
Methylene Blue	87-5733
Razor Blade	626931

	Carolina Biological Supply Catalog Number
A Leaf Elodea	16-2100
Lugol's Solution	87-2793
0.9% NaCl	88-8933
Muscle Tissue	312064
Cardiac	
Smooth	
Skeletal	
Multinucleate	
Human Blood (Wright Stain)	313158
Immersion Oil	85-2980
Nervous Tissue (Neurons) Stained with Silver Preparation	
Slide of Human Motor Neuron	313570

Exercise 8

"Fly-Nap"	17-3010
Drosophila Melangaster	17-2781
Vestigal	17-2841
Petri Dish	199279
Dissecting Microscope	59-1822
Hand Lens	60-2221
Paintbrush (Camel Hair)	

Exercise 9

500 ml *E. Coli*	21-1600
0.015 μg/ml pKAN	
Sterile Micropipel Tips	
Luria Broth, plates	
Luria Broth + Kanamycin plates	
Sterile Luria Broth Bottles	
95% Ethanol	86-1261
Sterile 15 ml Culture Tubes	
Spreading Rod	
0.5 – 10 μl Micropipet	21-4640
100 – 1,000 μl Micropipet	21-4644
Bunsen Burner	(several available)
Water Bath	216248
Permanent Marker	64-4298
250 ml Beaker	721502
Timer	69-6911
Disinfectant	19-9827

	Carolina Biological Supply Catalog Number
Pipetman	
Sterile Tip	21-5120
15 ml Culture Tubes	73-2097

Exercise 10

Cone Collection, dried	
Staminate Cone, preserved	
Alcohol/Gycerol	86-1710
Ovulate Cone, preserved	
Depression Slides	63-4200
Metric Ruler	70-2612
Monocot Flower, preserved	
Tobacco Dicot Flower, preserved	
Razor Blades	62-6931
Kodachrome Flower Slides	48-9900
Tradescantia Flowers (Anthers)	
Toothpicks	
Slides	63-2950
Slides (Pine Pollen)	301448
Slides—Male and Female Cones	301484
Cover Slip	63-2960
Distilled Water	85-7201
Seed Fruit Display	268848

Exercise 11

Tomato Seed	178670
Pots - 3″	665755
Sterile Vermiculite	
Solutions of Ca Deficiency	
Solutions of N Deficiency	
Solutions of K Deficiency	
Solutions of Mg Deficiency	
Solutions of P Deficiency	
Solutions of S Deficiency	
Solutions of Fe Deficiency	
Hoagland's Solution	
Flats with Capillary Matting	
Graph Paper	64-4620
Colored Pencils	64-4285

Exercise 12

Centrifuge	21-4075
pH Meter	(numerous available)

	Carolina Biological Supply Catalog Number
pH Paper	895511
Safety Goggles	GE09205
Gloves	(numerous available)
Test Tube	653310
Benedict's Solution	84-7111
Water Bath	
Urine Hydrometer (Urinometer)	72-2662
Cover Slip	63-2960
Slide	63-4200
Microscope (Binocular)	590959

Exercise 13

Prepared Slide (Testes)	316386
Microscope (Binocular)	590959
Immersion Oil	85-2980
Prepared Slide (Ovary)	316024
Graafian Follicle Slide	316012
Sea Urchin Embryology Kit	16-2505
Pipette	(various available)
Depression Slide	63-2200
Cover Slip	63-2960
Salt Water	
Prepared Slides of Starfish Development:	
Early and Late Cleavage	311126
Unfertilized through Gastrula	
Ovary	
Testis	
2, 4, 8, & 11 Blastomeres	
32 & 64 Blastomeres	
Composite showing major Stages	
Compound Microscope	
Monocular	595680
Binocular	590959
Prepared Slides of Frog Embryos	
Early Cleavage	309306
Late Cleavage	309306
Gastrula	309306
Late Yolk Plug	309306
Hatching Stage	309306
Chick Embryology Kit	309306
Incubator	309306
48-Hour Embryo	309306

	Carolina Biological Supply Catalog Number
33-Hour Embryo	309306
Egg	
Scissors	62-1775
Forceps	62-4784
Watch Glass	74-2368
Physiological Saline (Chick Ringers)	
Dissecting Needle or Probe	627420
Scalpel	62-5920
Dissecting Microscope	59-1822

Exercise 14

Greenhouse Flats	66-5670
Rye Grass Seeds	159313
Alfalfa or Clover Seeds	15-8282
Potting Soil	159705
Fertilizer	15-9765
2, 4-D	
Seed Starter Kit with Capillary Matting	
Gloves	(numerous available)
Goggles	GE09205
Liquid Fertilizer	
Scale	(numerous available)
Wheat Seeds	15-9395
Oat Seeds	

Exercise 15

Charts:	
Insect Sets	(various available)
Beetles	57-4514
Spiders, preserved	
Ticks, preserved	
Grasshoppers, preserved	
Leafhoppers, preserved	
Cicadas, preserved	
Aphids, preserved	
True Bugs, preserved	
May Flies, preserved	
Dragonflies, preserved	
Damsel Flies, preserved	
Dipter, preserved	
Hymenoptera (Bees), preserved	

	Carolina Biological Supply Catalog Number
Hymenoptera (Wasps), preserved	
Hymenoptera (Ants), preserved	
Lepidoptera (Moths), preserved	P920
Lepidoptera (Butterflies), preserved	
Coleoptera (Beetles), preserved	
Grubs (Beetle Larvae), preserved	
Dissecting Microscope	59-1822
Hand Lens	60-2221
Boll Weevil, preserved	
Tobacco Hornworm, preserved	
Flour Beetle, preserved	
Rosy Aphid	
Erwinia Tracheiphila (Micro-slide)	
Lady Bugs	
Tobacco Plant	
Tobacco Hornworm Eggs	
Bacillus Thuringiensis	15-4926
Plastic Plant Stakes	15-8982
Soil	15-9705
Pots - 3"	665774
Camel Hair Brush	
Diatomaceous Earth (Micro-slide)	295972
Diatomacious Earth	857570

Exercise 16

Mathematical Calculator	
Dissecting Microscope	59-1822
Hand Lens	60-2221
Simon & Schuster Field Guide to Insects	
Meter Sticks	70-2620
Golden Guide to Insects	45-4510
Collecting Shovel	
Labels	65-7480
Metal Rectangular Pan	62-9004
Forceps	62-4504
Small Glass Vials	716655
Gram Scale	702064
Drying Oven	
Simon & Schuster Field Guide to Trees	
Golden Guide to Trees	45-8117

	Carolina Biological Supply Catalog Number
Pruning Shears	
Collection Bags	
Notebook (field)	

Exercise 17

Safety Goggles	GE09205
Face Shields	646719
Antibacterial Soap Bactoshield	19-9827
pH Meter	
pH Paper	89-3930
Thermo Stick	
Conductivity Water Tester	
Conductivity and Dissolved Oxygen Meter	
Mercurial or Alcohol Thermometer	
Stop Watch	696911
Spectrophotometer	653302
Secchi Disk	65-2262
Beaker	71-7910
Water Bath	216248
Distilled Water	85-7201
LaMotte Kit	65-2450
Kim Wipes	63-3950
Magnetic Stirrer/Plate	

Exercise 20

Levowitz-Wever Stain Reagent	
Gram Stain Technique	82-1050
China Blue Lactose Agar Plates	821182
Safety Goggles	GE09205
Face Shield	646719
Gloves	(numerous available)
Antibacterial Soap	
Streptococcus Lactis Culture	15-5610
Enterobacter Aerogenes Culture	15-5030
Disposable Sterile Inoculating Loop	21-5850
Flat Slide	63-4200
Wooden Clothespin	
Bunsen Burner	706709
Wheaton Staining Jar	74-2150
Slide Holder	63-4397

		Carolina Biological Supply Catalog Number
Microscope		(various available)
Immersion Oil		85-2980
Petri Plates		74-1154
Antibiotic Disk	(Ampicillin)	80-6016
	(Erythromycin)	80-6136
	(Penicillin)	80-6296
	(Tetracycline)	80-6476
Autoclave Biohazard Bags		647054
Disinfectant		858475
Permanent Marker		
Sterile Swab		83-1920
Sterile Test Tube		19-9981A
Forceps		62-4540
Incubator		(numerous available)
Metric Ruler		70-2613
Sterile Water		798700
Sterile Petri Dish		199278
Safety Pipet Filler		736869
Pipet		(numerous available)
Water Bath		216248
Graph Paper		64-4620

EXERCISE 1

Measurements in Science: Units of Measurement

OBJECTIVES

Upon completion of this lab exercise, you should be able to:

- make measurements using a metric ruler and convert the metric units millimeters (mm), centimeters (cm), and meters (m), to the English units inches, feet, and yards.
- use a metric balance to obtain the weight of some common objects in the metric units gram (g) and kilogram (kg), and covert these weights to the English units ounce (oz) and pound (lb).
- use a flask and/or beaker to measure fluid volume in the metric units milliliter (ml), cubic centimeter (cc), and liter (1), and covert these fluid volumes to the English units of pint, quart, and gallon.
- measure temperature in degrees Celsius (°C) and convert temperature to degrees Fahrenheit (°F).

Suggested Reading:
Chapter 1 of *The Science of Agriculture: A Biological Approach*, 4th edition.

Note to Students:
A calculator is needed for this exercise. You might want to use a pencil to show your calculations.

INTRODUCTION

All scientists who do research or apply the findings of research need to take measurements or to use the measurements taken by other scientists. To facilitate this process, scientists around the world have developed a standardized scale of measurement, the **metric system**. The metric system is based on units that are multiples of the number ten and make converting from one unit to another simple. The United States was originally a British colony and we have long used the British system of measurement. We measure length in inches, feet, yards, and miles. We measure weight in ounces and pounds, and we measure volume using the units pint, quart, and gallon. Air, water, and soil temperature are currently measured on the Fahrenheit scale (°F), although many weather reports also give air temperature in Celsius (°C). The use of the English system of measurement is widespread in the United States and agriculture is no exception. For example, diesel fuel is priced per gallon (unit of volume). Grain is measured in bushels (unit of volume), and livestock weight is measured in pounds (unit of weight). However, products imported from Germany, France, Sweden, and Japan are built using metric measurements. If you buy a tractor built in one of these countries and you want to make repairs, you will need a metric set of wrenches to complete the job. Great Britain, Canada, and the United States are among the few countries in the world that do not employ the metric system. Most of us are comfortable using English units of measurement, but scientists need a standardized measurement system. Scientists need to be able to communicate with other scientists throughout the world, and as a result, scientific measurements are made using the metric system.

As students of agricultural science, you need to be familiar with the metric system. One day you may need to repair a Japanese tractor using metric tools, compare crop yield in the United States with crop yield in a European country, or apply a specific plant culture technique that was developed in another country. All of these tasks require a good working knowledge of the metric system. Let's begin by learning basic units of measurement. You need to remember only a single metric conversion for each unit of length, weight (mass), and volume. Some common conversion factors can be found in Table 1-1.

Conversion factors		
Length	**Mass**	**Volume**
2.54 centimeters = 1 inch	454 grams = 1 pound	16 fluid ounces = 1 pint
12 inches = 1 foot	2.2 pounds = 1 kilogram	2 pints = 1 quart
5,280 feet = 1 mile	1 ounce = 28 grams	4 quarts = 1 gallon
1 meter = 3.28083 feet		1 centimeter^3 = 1 milliliter
3 feet = 1 yard		1 teaspoon = 5 milliliters
1 yard = 0.9144 meter		1 fluid ounce = 30 milliliters
		1 gallon = 3.8 liters
		1 liter = 1.06 quarts

Table 1-1. Common conversion factors.

UNITS OF LENGTH

Length is generally defined as the distance between two points, end to end. The standard unit of length in the English system is the **inch**. Its counterpart in the metric system is a **centimeter**. One inch is equal to 2.54 centimeters (abbreviated *cm*) and three feet (equivalent to 1 yard) is equivalent to 0.9144 **meter** (abbreviated *m*). All metric conversions use multiples of 10 (see Table 1-2). Learn the metric prefixes which are based on the Latin terms for the units, such as centi- for one-hundredth, milli- for one-thousandth, and so on.

Prefix	Portion of unit (meter, liter, gram)
Kilo	1,000 or 10^3
Centi	1/100 or 10^{-2} (0.01)
Milli	1/1000 or 10^{-3} (0.001)
Micro	1/1,000,000 or 10^{-6} (0.000001)
Nano	1/1,000,000,000 or 10^{-9} (0.000000001)

Table 1-2. Some prefixes used in metric measurements.

The Solution:

$$10 \text{ miles} \times \frac{(5{,}280 \text{ ft})}{(1 \text{ mile})} \times \frac{(12 \text{ in})}{(1 \text{ ft})} \times \frac{(1 \text{ m})}{(100 \text{ cm})} \times \frac{(1 \text{ km})}{(1{,}000 \text{ m})} = \text{distance (km)}$$

If we cancel all units that appear in both the numerator (top) and the denominator (bottom) of the equation, we are left with the unit of interest, kilometers. This is one way to check to be certain that you have set up your equation correctly. We started with miles (our known distance) and ended with kilometers.

$$10 \; \cancel{\text{miles}} \times \frac{(5{,}280 \; \cancel{\text{ft}})}{(1 \; \cancel{\text{mile}})} \times \frac{(12 \; \cancel{\text{in}})}{(1 \; \cancel{\text{ft}})} \times \frac{(2.54 \; \cancel{\text{cm}})}{(1 \; \cancel{\text{in}})} \times \frac{(1 \; \cancel{\text{m}})}{(100 \; \cancel{\text{cm}})} \times \frac{(1 \text{ km})}{(1{,}000 \; \cancel{\text{m}})} = \text{distance (km)}$$

Next we multiply all the numbers in the numerator together, and multiply all of the numbers in the denominator together.

$$\frac{10 \times 5{,}280 \times 12 \times 2.54 \times 1 \times 1}{1 \times 1 \times 1 \times 100 \times 1{,}000} = \frac{1{,}609{,}344}{100{,}000} = \text{distance (km)}$$

The next step is to divide the numerator by the denominator.

$$\frac{1{,}609{,}344}{100{,}000} = 16{,}09344 \text{ km}$$

How do we convert English units to metric units of length? Suppose you want to know how far it is from your house to school in kilometers. You know that you live 10 miles from school and that 2.54 centimeters is equal to one inch. You also know that there are 100 centimeters in a meter and 1,000 meters in a **kilometer** (Table 1-2). You remember from math class that there are 12 inches in a **foot** and 5,280 feet in a mile.

The answer would be correctly expressed as 16.09 km since the most precise measurement is 2.54 cm/in. This value is two decimal places to the right, and our conversion is only as accurate as the most precise value. Notice that the answer has been rounded to two decimal places to the right of the decimal point. Since 0.00344 is less than 0.005, we round down to zero (recall from your earlier math experience the rules for rounding up and down).

Using the ruler provided by your teacher, measure the following objects using the English unit of length, the inch:

Length of your lab book = ___ inches
Length of your pencil or pen = ___ inches
Width of your desk top = ___ inches

Using conversion factors from Table 1-1, convert these values from the English unit of length, inches, to the metric unit of length, meters.

Length of your lab book = ___ meters
Length of your pencil or pen = ___ meters
Width of your desktop = ___ meters

Using the metric side of your ruler, measure three objects provided by your teacher using the metric unit of length, **millimeters** (abbreviated *mm*). Fill in the values in the second column in Table 1-3. Convert the length of these objects to feet. Write your answers in the third column of Table 1-3.

Name of object	Measurement (millimeters)	Measurement (feet)
1.		
2.		
3.		

Table 1-3. Measurements of three objects using the metric unit of length, millimeter, and the corresponding values in feet (English unit).

UNITS OF WEIGHT (MASS)

Mass is defined as the amount of matter in an object. We estimate mass by weight. We all have experience with the familiar English units of mass, the **ounce (oz)**, and the **pound** (lb). Less familiar are their metric counterparts, the **gram** (abbreviated *g*) and the **kilogram** (abbreviated *kg*). Referring to Table 1-2, we find that 1,000 grams is equivalent to 1 kilogram. Scientists such as physicists, chemists, and biologists measure mass using grams and kilograms. Many European, Latin American, and Asian countries also use the metric units of mass. Mass (weight) is typically measured using some type of scale. The scale you will be using to take measurements is similar to the scale used in many research labs and field stations where some degree of precision is needed, but not for very small quantities. The scale probably looks similar to the one in Figure 1-1. It is important to be certain that the scale is "zeroed" before you take the first value. To zero a **triple beam balance**, there is an adjustment or zero knob under the pan; turn the knob until the arrow on the right end of the arm is pointing to zero. It is important to be certain that the scale is zeroed so you do not over- or underestimate the mass of sample you are measuring.

Your teacher will give you a beaker filled with soil. It is your task to determine how much the soil weighs (in grams). When you place the beaker full of soil on the balance, you obtain the weight of the beaker and the soil together. How would you determine the weight of the soil assuming you cannot pour the soil out on the balance pan? You would estimate the weight of the soil by measuring the weight of the

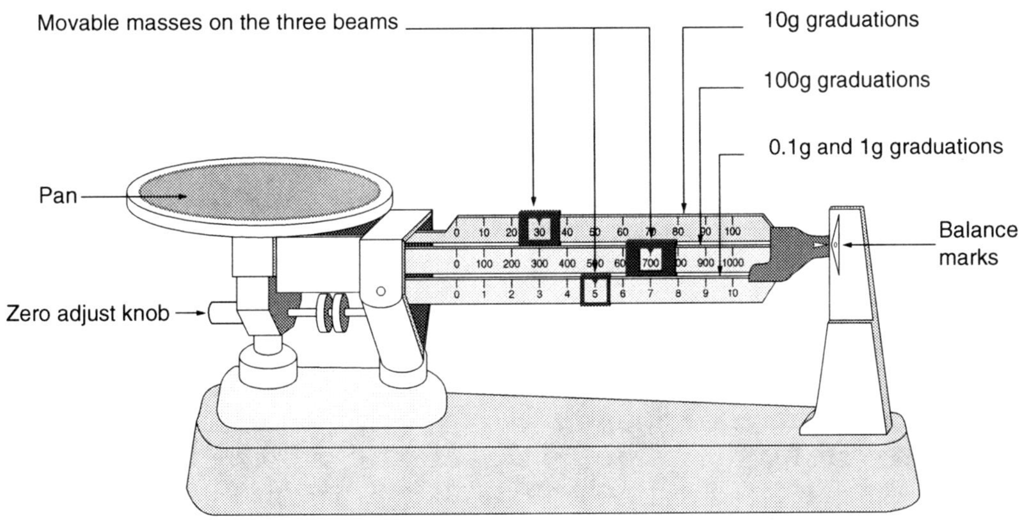

Figure 1-1. A triple beam balance.

Name of object	Weight/mass (in grams)	Weight/mass (in kilograms)	Weight/mass (in pounds)
1.			
2.			
3.			
4.			
5.			

Table 1-4. Measurements of five objects using the metric units of mass, gram, and kilogram, and the corresponding values in pounds (English unit).

beaker first and then measuring the weight of the beaker filled with soil. Subtract the weight of the beaker from the weight of the beaker plus the soil, and the weight of the soil remains. Scientists estimate weight (mass) by this subtraction method when a direct weight is not possible, for instance with liquid or gaseous substances. Three different objects will be given for you to weigh. Determine their weight in grams (g) using the balance. Convert their weight to kilograms (kg) and pounds (lbs) using the conversions from Table 1-1. Record your results in Table 1-4.

UNITS OF VOLUME

Volume is the amount of space occupied by a solid object or a liquid or gaseous substance. The standard English units of volume are pint, quart, and gallon. The metric units of volume are the **milliliter** (abbreviated *ml*), **cubic centimeter** (abbreviated *cc*), and **liter** (abbreviated *l*). Frequently, you may need to mix two substances together or dilute something by parts (nine parts water and one part liquid fertilizer, for instance). To work with liquids, the most frequent measurement is volume. Volumes measured on the English scale can be converted to the metric scale using the relationship: 1.06 quarts is equal to a liter. Cubic centimeters are frequently used in medical applications; medication for animals may be dosed in cc's per unit of body weight.

Several standard pieces of laboratory glassware are used to make volume measurements: the **graduated cylinder**, the flask (two types: *Erlenmeyer* and *Florence*), and the **beaker**. Figure 1-2 shows the relative

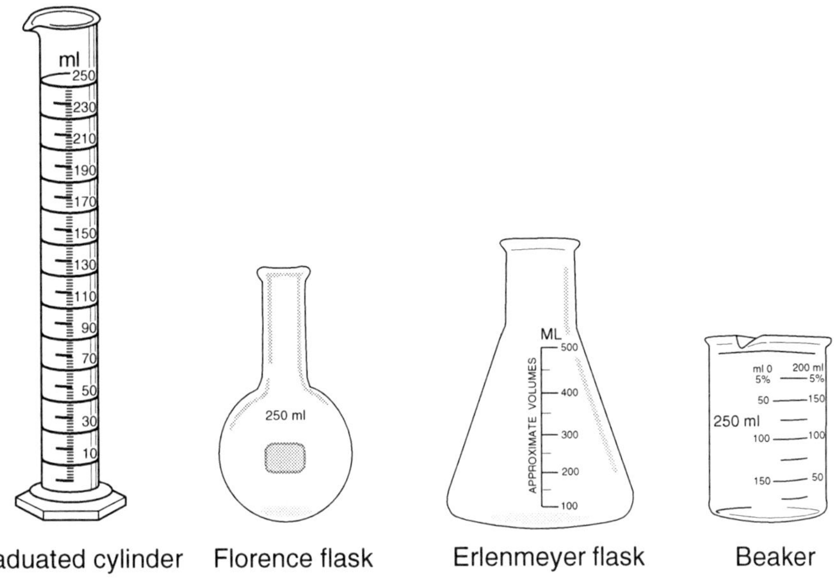

Figure 1-2. Types of general-purpose glassware used by scientists.

shapes of each of these pieces of glassware. The graduated cylinder is tall and thin with a wide circular base for stability. It yields the most accurate measurement of volume of the four types of glassware listed. It may be made of plastic or Pyrex (a substance that can withstand high heat).

The **Erlenmeyer flask** is cone-shaped with a flat base; it has a diverse range of use—from culture work to general lab use. The **Florence flask** is spherical with a rounded bottom, and it is generally used in the preparation of liquids that need to be concentrated by heating. Flasks are generally made of Pyrex. The rounded bottom gives a greater surface for evaporation. The beaker is the most common piece of glassware found in labs and field stations, and beakers can be made of plastic or Pyrex, depending on the intended use.

You will be using graduated cylinders of four different volumes: 10 ml, 50 ml, 100 ml, and 1,000 ml (1 liter) to make accurate measurements of liquid volume. To determine the volume of a liquid, look for the lowest place on the liquid surface when viewed from the side, the **meniscus**. The place where meniscus intersects the scale on the side of the cylinder is the volume. Say you are working in a greenhouse, and your supervisor has asked you to make a 10% solution of a liquid trace element fertilizer (by volume) to be added to young seedlings. How do you accomplish this task? Using the 100 ml graduated cylinder, pour 100 ml of the solution into a 1,000 ml (1 liter) Erlenmeyer flask. Then measure 900 ml of distilled water using the 1,000 ml (1 liter) graduated cylinder. Add 900 ml of distilled water to the 100 ml of trace element fertilizer, and you have the needed 10% solution. Using this principle and other glassware available (500 ml and 1,000 ml Erlenmeyer flasks; 50 ml, 250 ml, and 600 ml beakers), make

Ingredient	Metric value	English value
Whole milk	0.75 liter	
Sugar	500 grams	
Fresh strawberries	1 liter	
Lactobacillus cultures	45 grams	
Distilled water	50 milliliters	

Table 1-5. English equivalents for metric volume and weight values in a yogurt recipe.

the following solution: 500 ml of a 5% solution (use colored sugar water provided by your teacher).

Describe the relative volumes of sugar solution and water you mixed together to make the 500 ml 5% solution.

What did you notice about the color of the 5% solution as compared with the original solution?

You want to make fresh yogurt from milk, fruit, and active bacterial cultures, but your recipe is from a Danish dairy. You need to convert the quantities to English units. Fill in Table 1-5 with the correct volumes and weights needed to make your yogurt.

UNITS OF HEAT: TEMPERATURE

The metric unit most familiar to students is the unit of heat, the degree Celsius (°C). Many weather stations broadcast agricultural report temperatures in both Fahrenheit and Celsius. Temperature in °C are commonly displayed on banks and other signs around town. We know that water freezes at 0° and boils at 100°C. Similarly, water freezes at 32°F and boils at 212°F. What is the relationship between the **Celsius** and **Fahrenheit temperature scales**? Using the equations provided in Table 1-6, fill in the missing values in the table.

(1) °C = (°F − 32) × (5/9) Temperature °C	(2) °F = [(°C) − (9/5)] + 32 Temperature °F
40	
	98.6
	−15
	175
95	

Table 1-6. Equivalent values of temperature on the Celsius (°C) and Fahrenheit (°F) scales.

Use a thermometer with both a Celsius (°C) and Fahrenheit (°F) scale and estimate temperature for the following (let the thermometer equilibrate for 10 minutes before recording your temperature readings):

Temperature in the classroom = ___°C
Temperature in the classroom window = ___°F
Temperature in hallway outside of the classroom = ___°F
Temperature outside = ___°C

Questions for Thought

1. Why is it important for all scientists to use the same scale of measurement?

2. What are the advantages of using the metric system?

   The disadvantages?

3. Which system, the British system or the metric system, do you think will be in use by the year 2050? Why?

4. Using the metric scales of measurement, how are measures of volume and length related?

5. Is the distance between two degrees on the Fahrenheit scale greater or smaller than the distance between two degrees on the Celsius scale? Support your answer using actual numbers.

GLOSSARY OF TERMS

beaker: (*bikos*—earthen jug) A widemouthed cylindrical piece of glassware with a flat bottom used by scientists; may be made of plastic or heat-resistant glass.

Celsius temperature scale: Abbreviated °C, a temperature scale in which water boils at 100°C and freezes at 0°C.

centimeter: (*centum*—hundred) Abbreviated cm, a metric measurement of length, 10^{-2} meters.

cubic centimeter: Abbreviated cc, a metric measurement of volume based on a cube with 1 cm sides; also equivalent to 1 milliliter (mm).

Erlenmeyer flask: A cone-shaped, flat-bottomed piece of laboratory glassware; usually made of heat-resistant glass; named for Emil Erlenmeyer.

Fahrenheit temperature scale: Abbreviated °F, a temperature scale in which water boils at 212°F and freezes at 0°F.

Florence flask: A rounded piece of glassware with an elongated neck; may be flat-bottomed; named for the vessels in which wine was produced in Florence, Italy.

foot: Abbreviated ft or ′; basic unit of length in English-speaking countries; derived from the average length of the human foot, 12 inches.

graduated cylinder: A cylindrical piece of glassware with a metric volume measurement scale on the side used to measure fairly precise quantities of liquid; may be made of plastic or heat-resistant glass.

gram: (*gramme*—a small weight) Abbreviated g or gm, the basic metric unit of mass; equal to the mass of one cubic centimeter (milliliter) of water.

inch: Abbreviated in or ″, a unit of length in use in English-speaking countries equal to 1/36 of a yard; 1/12 of a foot.

kilogram: (*chilioi*—thousand, *gramme*—a small weight) Abbreviated kg, a metric measure of mass equal to 1,000 grams.

kilometer: (*chilioi*—thousand, *metron*—to measure) Abbreviated km, a metric measurement of length, 1,000 meters.

liter: (*litra*—a measure) Abbreviated l, basic metric measure of volume, equal to the volume of one kilogram of water at maximum density.

meniscus: (*mēniskos*—diminutive of moon, crescent) Relating to liquid; the curved portion of the liquid at the surface, concave when the walls of the cylinder are wetted and convex when they are not. The meniscus is the point at which volume is measured in a graduated cylinder or pipette.

meter: (*metron*—to measure) Abbreviated m, basic metric measure of length, equal to 39.37 inches.

metric system: (*metron*—to measure) A measurement scale based on multiples of the number 10. The basic units of measurement are: LENGTH-meter, MASS-gram, and VOLUME-liter.

milliliter: (*mitte*—thousandth) Abbreviated ml, metric unit of volume, 10^{-3} liters.

millimeter: (*mitte*—thousandth) Abbreviated mm, metric unit of length, 10^{-3} meters.

ounce: (*uncia*—twelfth part) Abbreviated oz, unit of mass based on the Roman pound and equal to 1/12 of a Roman pound.

pound: (*pondus*—weight) Abbreviated lb, unit of mass among English-speaking people equal to 16 ounces.

triple beam balance: A laboratory instrument used to weigh relatively small quantities of a substance.

EXERCISE 2

The Scientific Method: How Scientists Ask and Answer Questions

OBJECTIVES

Upon completion of this lab exercise, you should be able to:

- list the steps involved in the scientific method.
- write a hypothesis given a brief description of a scientific problem.
- identify the dependent and independent variables in a hypothesis statement.
- describe the purpose of a control in a scientific experiment.
- distinguish between quantitative and qualitative measurements.

Suggested Reading:
Chapter 1 of *The Science of Agriculture: A Biological Approach*, 4th edition.

INTRODUCTION

Most of the factual information contained in textbooks was the product of scientific inquiry. In many different disciplines, sociology, psychology, biology, physics, and agriculture, for example, investigators use the same process to ask and answer questions of interest. This process is called the **scientific method**. The scientific method is an important tool for researchers, who pose questions based on the curiosity they have of the world in which we live. However, it is also a useful model for problem-solving in everyday life. As students of agricultural science, you will encounter many problems in the field and in the livestock barn that will require investigation. You will need to be able to identify the problem, identify things that might contribute or cause the problem, and propose possible solutions. In order to become proficient at using this problem-solving method, you must first learn how it works. Let's explore the scientific method using an example from animal agriculture.

A catfish producer in Mississippi has noticed that the fish in one of his ponds are dying. When the fish from that pond are dressed, the meat is discolored and has an "off" odor. He notices a lot of "green stuff" floating on the surface of his pond. He wonders if this green stuff could be related to the death of his fish. The county extension office calls you to consult with the producer to try to find solutions for his problem, so he can salvage at least some of the fish in his pond.

Several steps are involved in the scientific method. The first step is the formulation of a **hypothesis**. A hypothesis is a statement of the problem, and it usually takes the form of a question. An example of a hypothesis that could be developed for the fish producer problem is:

Is fish death related to the amount of the green stuff on the surface of the pond?

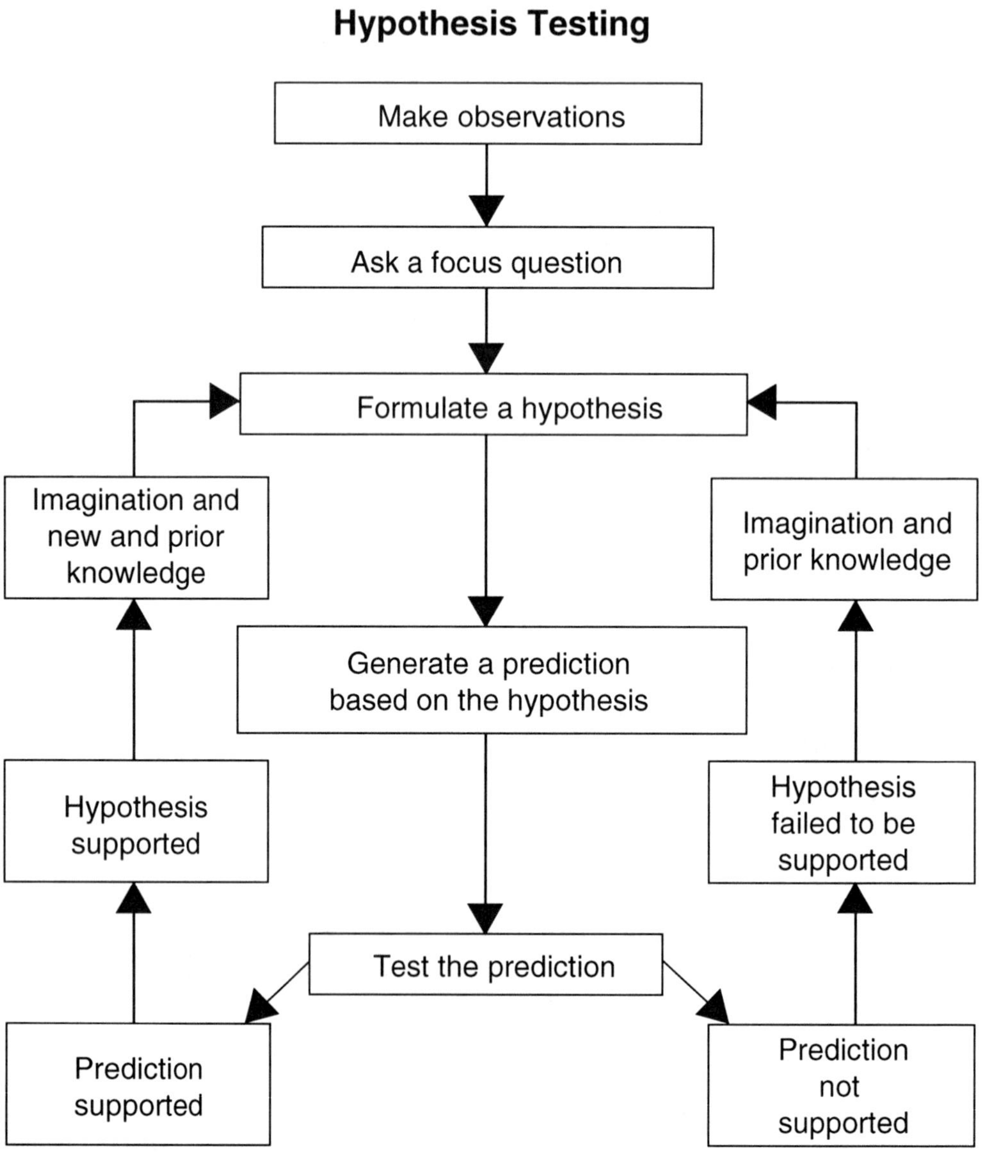

Figure 2-1. Steps of the scientific method.

A hypothesis should be testable, that is, you should be able to manipulate one of the parts of the question to get an answer. In the sample hypothesis, two parts or **variables** are given—fish death and amount of green stuff. Variables are factors that can exist in differing amounts. If the producer reduces or removes the green stuff from the pond surface, can we observe a change in the rate of fish death? The characteristic or variable that is manipulated, the presence or absence of green stuff in this example, is the **independent variable**. The rate of fish death, which changes with the amount of green stuff on the pond surface, is called the **dependent variable**. That is to say the number of dead fish is directly dependent on the amount of green stuff present on the pond surface. In simple problems, it is easy to identify the independent and dependent variables.

By following the formulation of the hypothesis and using all the available information, you can make predictions regarding the hypothesis. When you visit the affected pond, you notice a 3-inch thick mat of blue-green filamentous material. You suspect that this is some type of algae and ask the producer if you can take a sample of his "green stuff" to the local college for identification. A biologist at the college confirms your suspicion that it is a blue-green alga called *Anabaena*. The producer proposes that if the growth of the blue-green algae can be stopped or greatly reduced, his fish might have better survival rates. This prediction is related to the hypothesis and uses all of the available information. The next step in solving the producer's dilemma is to design an experiment to test the hypothesis. In the experiment, the value of the independent variable is changed, and its effect on the value of the dependent variable is observed. Let's design an experiment to test your hypothesis, remembering that the independent variable is the amount of blue-green algae and the dependent variable is the rate of fish death.

You obtain six large aquaria (all the same size), and you fill them with water taken from a healthy catfish pond at the producer's operation. You put each aquarium on the same bench in the laboratory where the light and temperature values are identical. You let the water stand for one day before starting the experiment. In Aquarium 1, you add 2 grams of algae (no fish), and in Aquarium 2, you place 20 small catfish fingerlings (no algae). In Aquarium 3, you add 20 small fingerlings and 2 grams of algae. In Aquarium 4, 20 small fingerlings and 4 grams of algae are added to the water. Aquarium 5 contains 20 small fingerlings and 8 grams of algae, and Aquarium 6 contains 16 grams of algae and 20 fingerlings. You make a note in your research notebook that you weighed each group of 20 fingerlings to be certain that their group weight was approximately the same. The aeration rate of each aquarium is identical. You make two observations at the same time each day for 2 weeks. This is the step of the scientific method that involves making observations and/or the collection of data. At the end of 2 weeks, your notebook data looks like Table 2-1.

Aquarium	Day 1	2	3	4	5	6	7	8	9	10	11	12	13	14
1. 2 grams algae														
2. 20 fish	0	0	0	0	0	0	0	0	0	0	0	0	0	0
3. 20 fish + 2 grams algae	0	0	0	0	0	0	0	0	0	0	0	0	2	1
4. 20 fish + 4 grams algae	0	0	0	0	0	0	0	0	2	0	0	4	0	0
5. 20 fish + 8 grams algae	0	0	0	3	0	5	0	0	0	2	0	0	0	0
6. 20 fish + 16 grams algae	0	3	0	5	0	8	4	0	0	0	0	0	0	0

Table 2-1. Fish death as a function of time and the amount of algae present.

The numbers in each column represent fish death. From the data you collect, it is obvious that as the amount of algae increases, the rate of fish death increases. You notice also that as the amount of algae present increases, the fish die earlier. You share your results with the producer who asks why you did not put any fish in the first aquarium or any algae in the second aquarium. You tell the producer that these two aquariums are **controls** for the experiment. The first aquarium allows you to observe algal growth in the pond water. The second aquarium would give you an estimate of fish mortality in the absence of any algae. This is important because the fish could have been dying in response to something else in the pond water. Because no fish in Aquarium 2 died, you conclude that the fish death was attributable to the *Anabaena*. You have analyzed your results with respect to the original hypothesis. The data collected support the hypothesis, but the experiment does not provide you or the producer with any solutions. So you suggest to the producer that you look at some additional data that you collected while conducting the experiment. You suggest that the algae may be using all of the available oxygen dissolved in the water, and the fish are suffocating. Many scientific investigations yield additional questions for further study.

In some experiments, not all of the variables have numerical values. In our example of the fish producer with the algal bloom, recall that the producer reported that the dressed fish from the affected pond had an "off odor." The assessment of odor and taste are important to fish producers; it determines the value of the crop. Odor and taste do not have underlying numeric value; the observations of taste and odor are **qualitative measurements**. We can determine which fish smell better or worse than a given standard, and similarly we can compare the taste of cooked fish. The qualities of odor and of taste are our variables of interest in this case. In contrast, the rate of fish death is a variable for which real numerical values exist. We make **quantitative measurements** regarding fish mortality and/or dissolved oxygen—these are quantitative variables.

Now that you have had a thorough introduction to the scientific method, it is time to practice applying what you have learned. The description of several agricultural problems follows. For each problem, identify the independent and dependent variables, formulate a hypothesis, and design an experiment to test the hypothesis. Remember that making qualitative measurements is more subjective than making quantitative measurements.

Problem 1

You are raising hogs for market, and your veterinarian recommends that you switch the type of feed given to the mature hogs. The veterinarian is concerned that the present feed is too high in protein. Although a high protein diet is recommended for growing juveniles, food too high in protein can cause kidney problems in adult animals. You switch feed and notice that the weights of your mature animals drop. You want healthy animals with maximum weight, but you do not know how to solve the problem. You call the agriculture teacher for help.

Write a hypothesis that describes the problem.

Identify the independent variable.

Identify the dependent variable.

Are the variables quantitative or qualitative? Explain your answer.

Design an experiment to test the hypothesis you have written.

Problem 2

You have planted soybeans in a field with clay soil. The field is predominantly flat with a slight slope at one end where a creek borders the field. You notice that germination and growth is slowest in the flattest portion of the field. Conversely, you also notice that you got good germination and rapid growth on the slight hill that meets the creek. You wonder why there should be differences in growth and germination in different parts of the field. Is it due to the extra water near the creek? Is there something different about the soil in that portion of the field, or is it due to the difference in slope or site orientation (north, south, east, west). Baffled, you call the agronomist at the nearest experimental station in an effort to salvage your crop.

Write a hypothesis that describes the problem.

Identify the independent variable(s).

Identify the dependent variable(s).

Are the variables quantitative or qualitative? Explain your answer.

Design an experiment to test the hypothesis you have written.

Problem 3

You have recently purchased a small farm, and the previous owners provided you with copies of their crop yields and planting schedules. The growing season is long enough for two crops per year in each field. You notice that in two fields, winter wheat was alternated with corn as a summer crop. In the other three fields, winter wheat was alternated with soybeans as a summer crop. Initially, the winter wheat yield/acre in all fields was about the same. After 4 years, the winter wheat yields in the fields containing summer corn crops were lower than in the fields containing soybean crops. Following 2 years of poor wheat and corn harvests, the application of a high nitrogen fertilizer in those fields brought crop yields close to those of the soybean/wheat fields. You are puzzled by the differences in crop yield and fertilizer application rates in the corn/wheat fields as compared with the soybean/wheat fields.

Write a hypothesis that describes the problem.

Identify the independent variable(s).

Identify the dependent variable(s).

Are the variables quantitative or qualitative? Explain your answer.

Design an experiment to test the hypothesis you have written.

Questions for Thought

1. List the steps in the scientific method.

2. Is the scientific method just for "scientists"? Why is it important that a nonscience student understand the scientific method? How do you think that this method can be used in everyday life?

3. Distinguish between independent and dependent variables. Give an example of each.

4. What is the purpose of a control in an experiment?

5. Differentiate between qualitative and quantitative measurements. Under what circumstances would you make qualitative measurements/observations? Give an example of an experimental situation in which quantitative measurements might be taken.

GLOSSARY OF TERMS

control: The portion of the experimental population which does not receive the treatment; all variables are constant. Results from the experimental groups are compared for each variable with the measurements from the control group for those variables to determine if differences exist as a result of the experimental "treatment."

dependent variable: The factor that changes in response to the manipulated (independent) variable.

hypothesis: (*hypo*—under, less; *tithenai*—to put) A conjecture which is based on previous observation(s) and which serves as the basis for further scientific speculation and/or experimentation.

independent variable: The factor that is changed or manipulated by the experimenter.

qualitative measurements: Measurements and/or observations in a scientific experiment which describe a property or characteristic of a study subject/object. Qualitative measurements do not involve the assignment of numbers in observations but rather the assignment of qualities such as color, odor, gender, etc.

quantitative measurements: Measurements and/or observations in a scientific experiment which involve the assignment of a true numerical value, for example: temperature, weight, growth rate, etc.

scientific method: A systematic method for examining, recognizing, and formulating scientific problems; designing experiments, collecting data, and evaluating the results of experiments to further refine one's understanding of the initial problem.

variable: A characteristic in an experiment which assumes more than one value; for example, temperature is a variable if measurements are made at different temperatures such as 0°C, 10°C, 25°C, etc.

EXERCISE 3

Measurements in Science: Microscopes

OBJECTIVES

Upon completion of this lab exercise, you should be able to:

- name the parts of the compound microscope and give their function.
- name the parts of the dissecting microscope and give their function.
- distinguish between parfocal and parcentric.
- calculate the total magnification power of a microscope given the magnification of the ocular and objective lenses.
- determine whether to use a compound microscope or a dissecting microscope to get the best view of the specimen.
- distinguish between a transmission electron microscope and a scanning electron microscope.

Suggested Reading:

Chapter 1 of *The Science of Agriculture: A Biological Approach*, 4th edition.

Note to Students:

A calculator is needed for this exercise. You might want to use a pencil to show your calculations.

INTRODUCTION

All scientists make use of specialized tools to make observations. Depending on the scientific discipline, scientists may use a variety of machines or tools. Scientists may use balances to weigh chemicals or living organisms, glassware for chemical reactions, instruments that analyze the chemical composition of substances, computers to store data or analyze results, or specialized microscopes to view objects. Each scientific area makes use of many different tools, and as students of agricultural science, you will make use of many of them. In this exercise, you will familiarize yourself with several different types of microscopes including their use and limitations.

Early scientists were limited by the resolving power of the human eye; they could only make observations of organisms and nonliving specimens that were 1 mm in size or larger. It became obvious that in order to explore the structure of plants and animals more closely, some type of magnification was necessary. In 1677, a Dutch scientist named Anton von Leeuwenhoek made the first primitive light microscope. With his invention, Leeuwenhoek examined pond water, human blood cells, and a variety of other specimens. The small, moving organisms he observed in the pond water amazed him. No further progress was made until the English scientist Robert Hooke used a device similar to Leeuwenhoek's to view cork. Cork is composed of dead plant cells; only the walls surrounding the cells remain. Hooke looked at the regular structures and thought they resembled tiny compartments (rooms) or cells, similar to the small spaces in monasteries occupied by monks. As the study of living things evolved, the light microscope was further refined. With the better light microscopes currently available, scientists can look at objects that are as small as 1 μm (10^{-6} m or 0.00003937 inches) (see Figure 3-1). To look at even

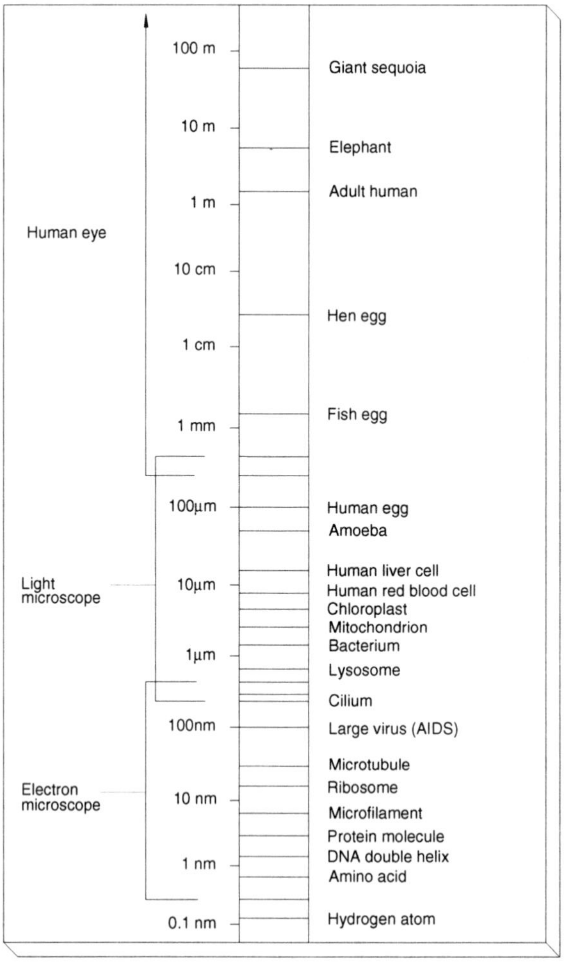

Figure 3-1. Relative size of specimens commonly encountered in the life sciences.

smaller specimens, special microscopes that have an electron beam instead of a light beam to illuminate the object. These microscopes are called electron microscopes. One disadvantage of the electron microscopes is that the object being viewed is placed in a vacuum. Living organisms cannot survive in a vacuum; therefore, only dead organisms or nonliving objects can be viewed using an electron microscope. Organisms or objects that are sliced into thin sections are viewed with a **transmission electron microscope**. Disease-causing bacteria and viruses in plant and animal cells are frequently viewed using a transmission electron microscope (see Figure 3-2). Sometimes, the surface of a small specimen is of interest, and a **scanning electron microscope** is used. We have all seen pictures of a fly's eye, butterfly's wing, or a pollen grain taken with a scanning electron microscope (see Figure 3-3).

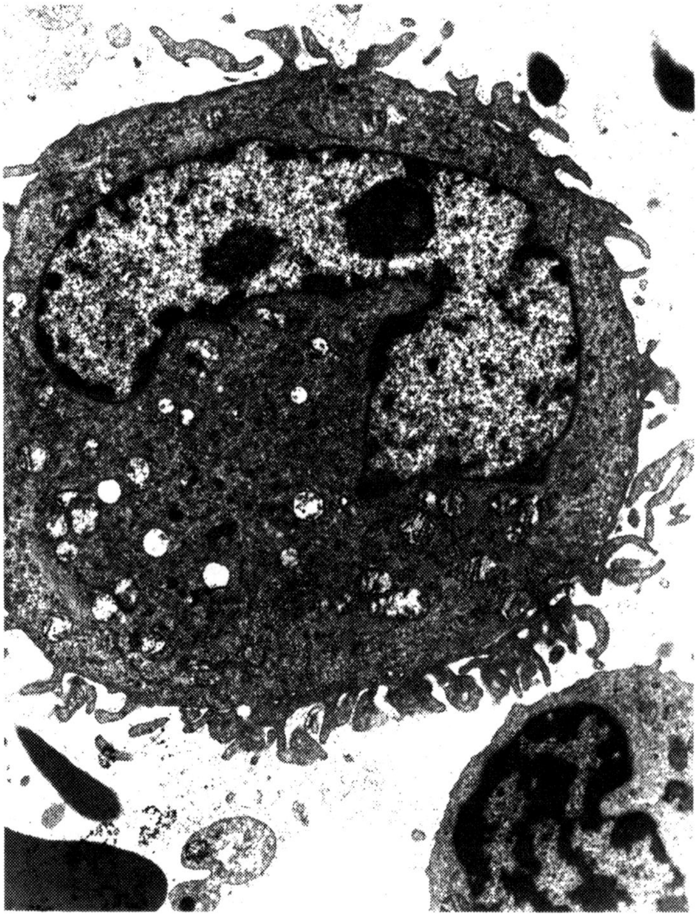

Figure 3-2. Canine white blood cell magnified 20,000x, taken using a transmission electron microscope. Courtesy Dr. Walstine L. Steffens, University of Georgia College of Veterinary Medicine.

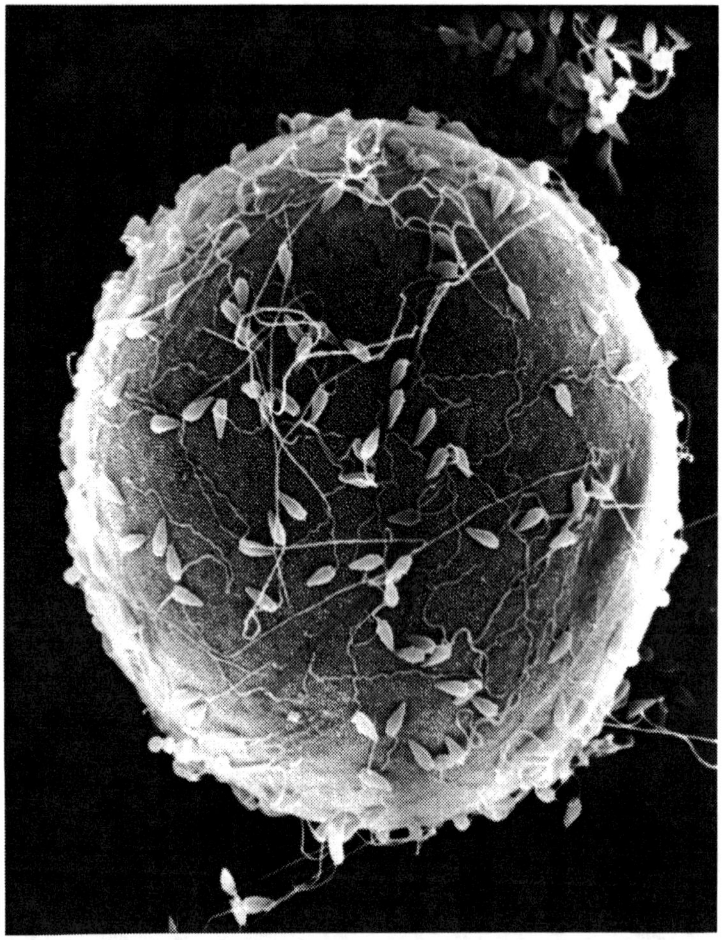

Figure 3-3. Sea urchin egg fertilization magnified 200x, taken using a scanning electron microscope. Courtesy Dr. Walstine L. Steffens, University of Georgia College of Veterinary Medicine.

LIGHT MICROSCOPES

The type of microscope most familiar to the student of the applied life sciences is the one used in most biology courses, the light microscope. In contrast to electron microscopes, light microscopes allow the viewer to make observations of living cells and cell activities. There are two types of light microscopes: the **compound microscope** and the **dissecting (or stereo) microscope**. The differences between these two types of light microscopes are based on the **magnification** and **resolution** of each kind of microscope. Magnification is the extent to which the object's size is increased. Resolution is the smallest distance between two points that can still be viewed separately. The compound microscope has greater magnifying power and resolution than the dissecting microscope.

COMPOUND MICROSCOPE

The compound microscope has at least two lenses, the **ocular** and the **objective** lenses. The ocular is the lens at the top or front of the microscope where you place your eye. Microscopes that have two oculars (eyepieces) are called binocular microscopes in contrast to mon-ocular microscopes that have a single eyepiece. The objective lens is located at the end of the body tube containing the reflective mirrors. Most compound microscopes have several objective lenses of varying magnification (see Figure 3-4). Most compound microscopes have several objective lenses. The lens shortest in length is the low-power objective lens and usually has a magnification of 4× or 10× depending on the microscope. Microscopes that lack a condenser will usually have a high-power objective lens with a magnification of 40×. The total magnification of the specimen is greatest when this lens is in use. The **condenser** with an **iris diaphragm** is found in some compound light microscopes. The condenser and iris diaphragm focus light evenly along all points of the specimen on the microscope slide. This increases the contrast so that the resolution of the microscope is increased.

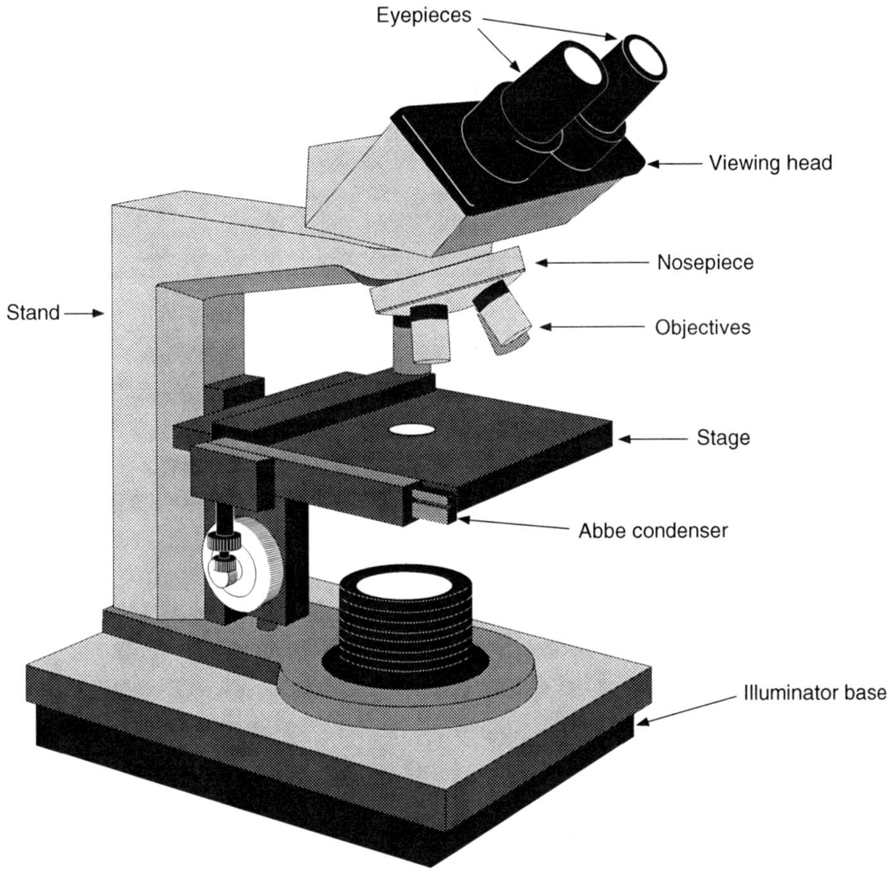

Figure 3-4. Compound light microscope.

Microscopes that use oil immersion objective lenses also have a condenser with an iris diaphragm. The **condenser** has an **adjustment knob** that allows the user to change the distance between the condenser and the bottom of the microscope slide. The iris diaphragm is like the aperture on a camera; by opening and closing the diaphragm, the microscope user can regulate the amount of light that passes through the slide/specimen. The high-power objective lens—the oil immersion lens (usually 100× magnification)—is used by placing a drop of special oil (called immersion oil) on the slide, adjusting the condenser so that it is as close to the microscope slide as possible, and adjusting the iris diaphragm to allow concentrated light to pass through. The oil immersion lens is used to maximize the clarity of visible details or to view very tiny organisms such as bacteria. The immersion oil bends the light rays so that they provide maximum resolution with high magnification.

How do you determine the total magnifying power of a microscope?

Ocular × Objective = Total Magnification

Examine your microscope. Calculate the total magnification available with each objective lens by completing the following table.

Ocular	× Objective	= Total Magnification
OCULAR	LOW-POWER OBJECTIVE	=
OCULAR	MEDIUM-POWER OBJECTIVE	=
OCULAR	HIGH-POWER (DRY) OBJECTIVE	=
OCULAR	OIL IMMERSION OBJECTIVE	=

Obtain a slide of the letter *e* from your teacher. Orient the slide so that the letter *e* appears right side up (correctly oriented for reading). Place the slide on the center of the **stage** (flat table portion under the revolving nosepiece). Try to position the slide so that the light coming through the stage illuminates your letter *e* (this means that the specimen—*e*—is properly centered on the stage). Rotate the nosepiece so that the low-power objective lens is in place. While looking through the ocular(s), rotate the **coarse adjustment knob** (the largest knob on the side of the microscope) until your letter *e* comes into view. Make your best focus with the coarse adjustment knob, and then use the **fine adjustment knob** (the smaller knob on the side—see Figure 3-4) to make the final focus.

When looking through the microscope, how does the appearance of the letter *e* compare with the orientation on the slide?

Move the slide to the right while looking through the ocular. In what direction does the slide move?

You need to remember when working with any compound light microscope that the arrangement of lenses results in an image that is upside-down and backwards from its orientation on the slide. If you are looking through the microscope, and you wish to move the specimen to the right, you will need to move the slide to the left. It will take some practice before you are accustomed to this compound microscope quirk. Now that you are in focus on lower power, move the revolving nosepiece so that the next objective lens is in place. You should find that the *e* is still in focus. This compound light microscope is **parfocal**, that is, when an object is in focus using the low-power objective lens and the revolving nosepiece is moved to a higher power objective, the object should remain in focus. This feature ensures that properly adjusted compound light microscopes have focal points in the same planes for all objective lenses. Similarly, if your letter *e* is in the center of your field of view (the circle of light you see when you look through the ocular lenses), it should remain centered when you change to another objective lens. The compound light microscope is also **parcentric**.

Depth of field is another principle of microscope use. This is the working distance between the objective lens and the specimen. Compound and dissecting microscopes have differing depths of field; the greater working distance is found in the dissecting microscope. You must also remember that the microscope slide is three-dimensional. To demonstrate depth of field, you will need to obtain a slide with three different-colored threads on it. These threads cross at different places on the slide, and the slide preparation is relatively thick. Use extra care in using this slide with the longer, higher power, objective lenses. Begin by placing the colored slide on the stage so that the thread intersection is in the middle of the field of view (circle of light). Using the low-power objective and the coarse adjustment knob, bring the colored threads into focus. Make any necessary adjustments in the position of the threads on the stage. Move the revolving nosepiece until the high-power objective is in place. Using the fine adjustment knob, bring the colored threads into focus. *Remember, the coarse adjustment knob is not used with the high-power or oil-immersion lenses!* While looking through the ocular and moving the fine adjustment knob, you should see several different thread colors come in and out of focus. This illustrates depth of field.

Which thread color is on top?

Which thread color is in the middle?

Which thread color is on the bottom?

There is an inverse relationship between the size of the field of view and total magnification. That is, as magnification increases, the size of the field of view decreases. To investigate this relationship, obtain a clear plastic metric ruler from your teacher and clip it down on the stage using the stage clips. Try to center the metric scale in the middle of the field of view. Focus on the ruler using the low-power objective lens. How many millimeters (mm) wide is the field of view? Record your measurements in the table below. (*Hint:* Remember that total magnification = ocular × objective.)

Total magnification	Width of field of view (mm)
Low power	
Medium power	
High power-dry	

Switch to the next objective lens and focus again on the metric scale on the edge of the ruler. How many millimeters wide is the field of view?

Record your measurement in the table above. Has the field size increased or decreased?

Switch to the high-power objective lens. Are any divisions on the edge of the metric scale visible?

How wide is the field of view? Record your measurement in the table.

The inverse relationship between field size and magnification frustrates many new microscope users. "I've lost it!" is a frequently heard complaint. In reality, nothing is lost; the specimen was not in the center of the field of view, and as the field size decreased as magnification increased, the specimen was no longer in the field. A quick adjustment of slide position can usually recover the lost item. Now that you have become an expert with the compound microscope and prepared slides, let's practice your newfound skills with living material.

MAKING A WET MOUNT

Living material for observation must be protected from the heat generated by the concentrated light source; the specimen is usually suspended in water or some other liquid—a "wet" mount. Obtain a clean slide and cover slip (sometimes called a cover glass). Clean the ocular and all of the objective lenses with lens paper before you begin and after you have finished. This is a habit that will keep your microscope in good working order. Place the slide on a flat surface and add a small drop of pond water or some other living culture on the center of the slide. Take the cover slip and place it on its edge so that it contacts the drop of liquid at an angle (see Figure 3-5). When the edge of the drop of liquid contacts one side of the cover slip, drop the cover slip onto the liquid. It should

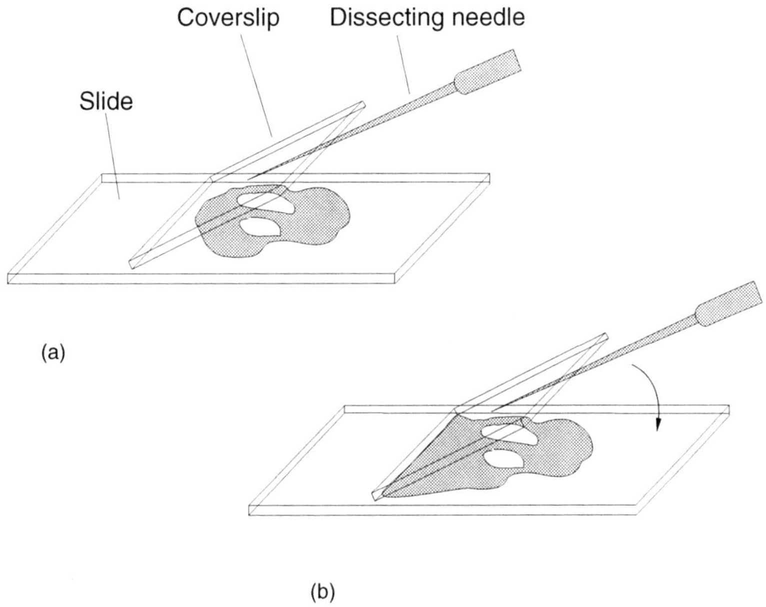

Figure 3-5. Procedure for making a wet mount.

flatten out with a minimal number of air bubbles. Place the slide on the stage and begin by focusing using the low-power objective. If you see any clear circular objects outlined in black, these are air bubbles. Gently tapping the surface of the slide with a pencil eraser should force any air bubbles out of the edge of the cover slip. Once your slide is in focus under low power, switch to the next objective, using the fine adjustment knob to make your focus. Draw what you see in the space provided.

DISSECTING MICROSOPE

The dissecting microscope (sometimes called a stereo microscope) has a greater working distance and depth of field than the compound light microscope (see Figure 3-6). The object can be illuminated from the bottom (as with the compound microscope) or from the top. The total magnifying power of the dissecting microscope is far less than that of the oil immersion compound light microscope.

Because of this, dissecting microscopes are used to view surface details of larger objects, dissect small specimens (such as chicken eggs, insects, or flowers), or to view larger living specimens such as the flatworm, earthworm, or leech.

Your teacher has a variety of different objects, living and nonliving, for you to use with the dissecting microscope. Draw and label three different specimens in the space provided.

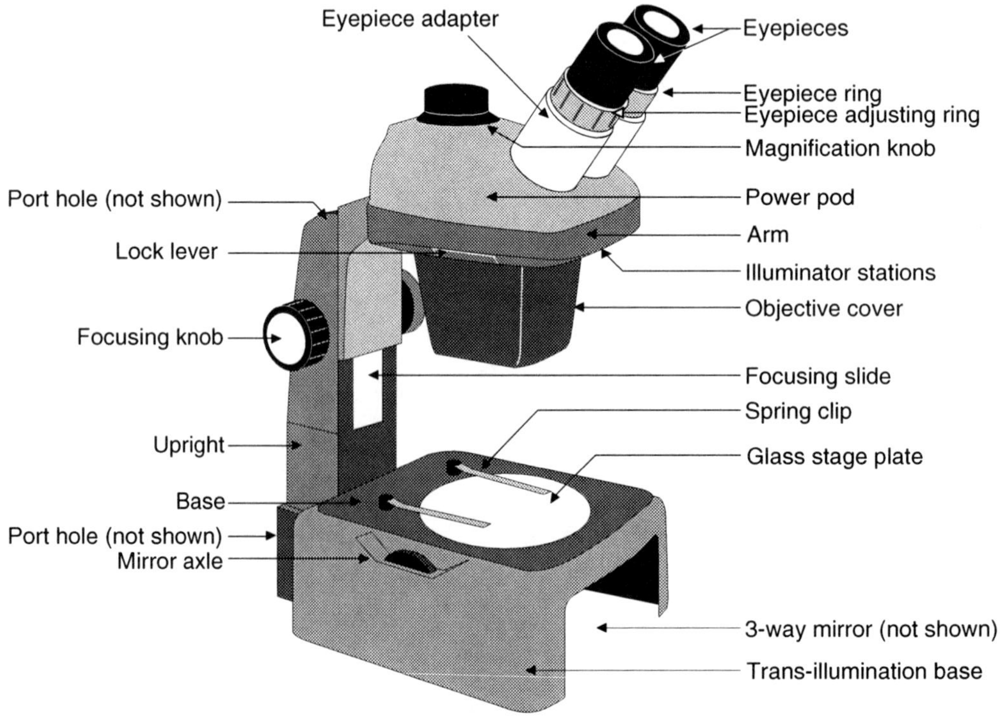

Figure 3-6. The dissecting (or stereo) microscope.

Questions for Thought

1. Distinguish between parfocal and parcentric. How does the parfocal and parcentric nature of the compound light microscope affect its use?

2. Describe how the total magnifying power of a microscope is calculated.

3. Under what circumstances would you use a compound light microscope?

4. Under what circumstances would you use a dissecting microscope?

5. Describe the relationship between field of view size and total magnification. How does this affect microscope use?

6. Distinguish between the scanning electron microscope and the transmission electron microscope.

7. Why is the microscope such an important scientific tool? How different do you think life would be without the microscope?

GLOSSARY OF TERMS

coarse adjustment knob: Large outside adjustment knobs located on each side of the microscope at the base (see Figure 3-4); used to make initial focus of the specimen. The coarse adjustment knob is to be used with the *4× and 10× objectives only.*

compound microscope: A microscope that has both an ocular and an objective lens mounted in a body tube; uses light to illuminate the viewing specimen; specimen is illuminated from the bottom. With powerful objective lenses, very small specimens (such as bacteria and sperm) can be viewed.

condenser adjustment knob: Located below the stage on the left side of the microscope; regulates the distance of the condenser from the stage and assists in focusing the light source on the specimen through the condenser lens (see Figure 3-4).

condenser iris diaphragm: Portion of the microscope which regulates the amount of light passing through the specimen (see Figure 3-4).

dissecting microscope: Microscope with a larger working distance between the specimen being viewed and the objective lens. Practically, this larger working distance means that the specimen can be illuminated from above and below. The total magnifying power of the dissecting microscope is less than that of the compound microscope, and its use is limited to viewing details of larger objects.

fine adjustment knob: Small inside adjustment knobs located on each side of the microscope at the base (see Figure 3-4), used to make final focus of the specimen. The fine adjustment knob is the *only adjustment knob to be used with the 40× and 100× objectives.*

magnification: The appearance of enlargement by a microscope; it is determined by multiplying the magnifying power of the ocular lens (eyepiece) by the magnifying power of the objective lens.

objective: Mounted on the revolving nosepiece, the objectives contain lenses of differing magnification and are used in conjunction with the ocular lenses to view the specimen. Most microscopes have several different objectives: *4×—low-power objective; 10×—medium-power objective; 40×—high-power objective;* and *100×—oil-immersion objective.*

ocular: Eyepiece(s) containing a lens located at the top/front of the microscope; used to magnify the specimen (see Figures 3-3 and 3-4). Microscopes with a single eyepiece are *monocular* microscopes, while microscopes with two oculars are *binocular* microscopes.

parcentric: (*par*—equal; *centrum*—center of a circle) The property of compound light microscopes in which a specimen appears in the center of the field of view when observed using the lower power objective remains in the center of the field of view when a higher power objective lens is used.

parfocal: (*par*—equal; *fokus*—hearth) The property of having lenses with corresponding focal points in the same plane. In practice, a parfocal microscope allows the user to focus on the specimen using the 10× objective, and when the user switches to the 40× objective, the specimen should remain in focus.

resolution: The smallest distance between two points which can still be viewed separately.

scanning electron microscope: A microscope that passes an electron beam across the surface of the specimen being viewed; the specimen has been coated with a thin layer of gold, and the electron beam excites the electrons on the gold surface. A tube similar to those used in televisions picks up the signals transmitted by the electrons, and a picture of the specimen's surface is produced (see Figure 3-3).

stage: The flat horizontal portion of the microscope on which the specimen is placed for viewing (see Figure 3-4).

transmission electron microscope: A microscope in which an electron beam is passed through specimens which have been stained with electron-dense heavy metals. The resulting photographic image is called a *photomicrograph* (see Figure 3-2).

EXERCISE 4

Soils: Physical Properties of Soil

OBJECTIVES

Upon completion of this lab exercise, you should be able to:

- distinguish between the organic and inorganic components of soils.
- determine the pH of the soil and relate pH to the soil's parent material.
- distinguish between gravel, sand, loam, silt, and clay and relate these terms to relative particle sizes in a soil sample.
- relate water saturation rates to soil texture.
- make recommendations about soil tillage based on the soil's physical characteristics, slope, and climate.

Suggested Reading:
Chapter 2 of *The Science of Agriculture: A Biological Approach,* 4th edition

INTRODUCTION

Soils provide the basis for all terrestrial plant growth on earth. Soils are complex systems derived from the activities of climate, geologic events, plants, and animals (including humans). All soils are composed of two types of material: organic and inorganic material. **Organic** materials contain both the elements carbon (C) and hydrogen (H). Organic materials in soil are created from broken down plant and animal materials, and are generally concentrated near the surface of the soil, comprising between 3% and 8% of the **topsoil** (McRae, 1988). Fresh organic matter provides a substrate for microbial growth. Soil microorganisms decompose fresh organic matter forming a more stable substance called **humus**. In the **subsoil**, organic matter comprises less than 1% to 2% of the soil. The mineral portion of the soil is **inorganic** (it lacks the elements carbon and hydrogen) and is derived from the parent material (bedrock). The physical and chemical properties of the organic and inorganic soil layers define soil type and determine its suitability for agricultural practices.

Soil **pH** is a measure of the relative concentration of hydrogen ions (H+). Soils that have high concentrations of hydrogen ions are acidic, and soils that have low concentrations of hydrogen ions are alkaline (or basic). Soil pH is important because it affects plant growth. In acidic soils high in metals such as iron and aluminum, plant uptake of these metals increases as pH decreases. At a point, these metal ions become toxic to the plants, and the soil pH must be adjusted before it becomes suitable for plant cultivation. Soil pH is related to the amount of organic material in the soil and to the pH of the rock from which the soil is derived. For example, limestone produces alkaline soils in contrast to granite, which produces acidic soils.

One method for soil classification involves characterizing soil texture, the relative proportion of mineral particles of different sizes. There are four different size particles in the USDA System of soil classification: gravel, sand, silt, and clay. It should be noted that an alternative system of classification, the International System, is in use in many other parts of the world and uses slightly different criteria than the USDA System.

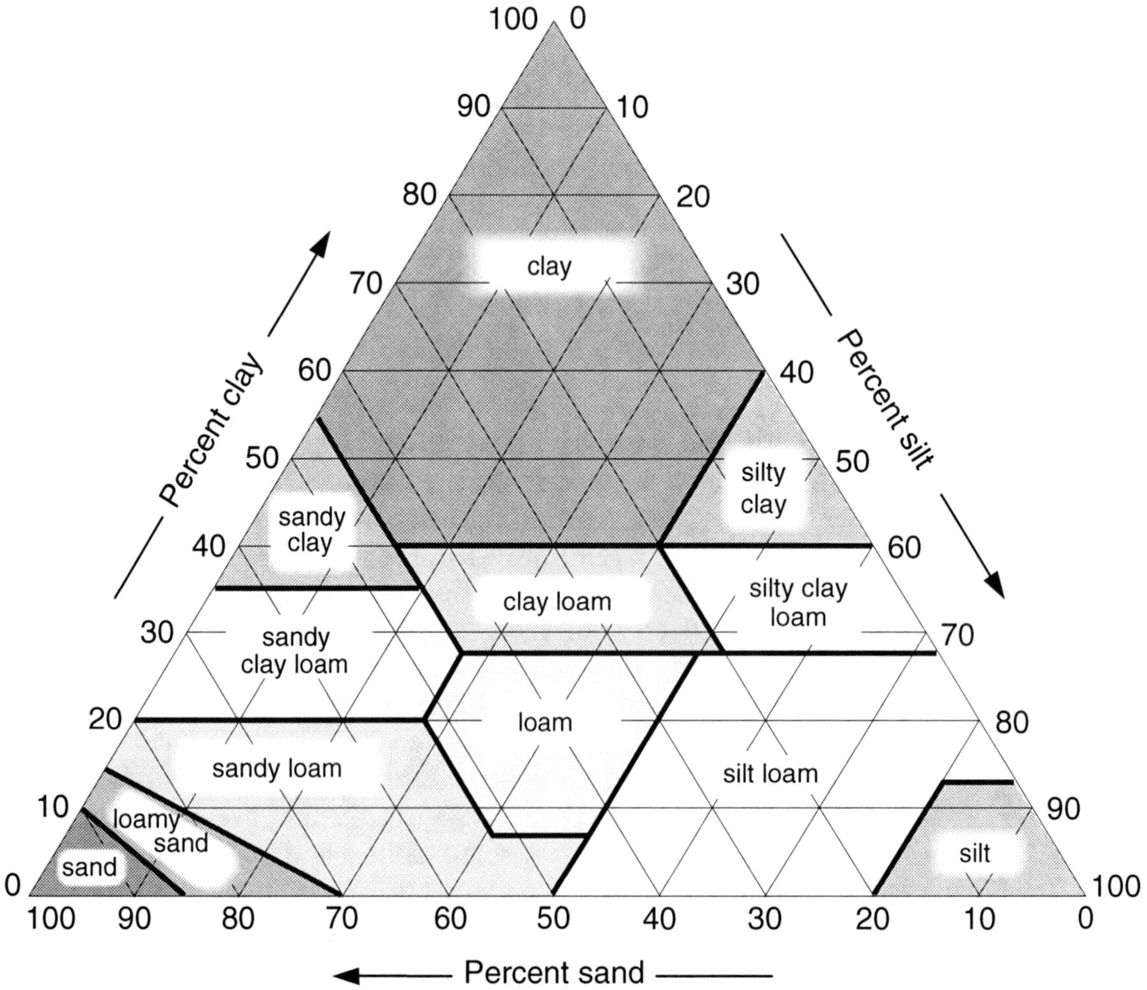

Figure 4-1. Relationship between soil particle size distribution and soil texture class.

Gravel is defined as a mineral particle that is larger than 2.00 mm in diameter. **Sand** falls within a range of 2.00 to 0.05 mm in diameter, while **silt** particles are between 0.05 and 0.002 mm in diameter. **Clay** particles are less than 0.002 mm in diameter. **Loam**, an aggregate of different-sized particles, is composed of less than 52% sand, 28% to 50% silt, and 7% to 27% clay (see Figure 4-1). Loam is the most desirable soil class for cultivation.

MEASURING SOIL pH

Because soil pH is one of the physical properties that determines how well plants grow, it is important to be able to measure soil pH. The method you will use involves measuring the pH of soil suspended in distilled water. Your teacher will give you three soil samples: Sample A: cultivated field with nitrogen fertilizer added recently; Sample B: forest soil; and Sample C: soil from an undisturbed fencerow or pasture.

Hypothesis

Write a hypothesis that relates soil pH to the type of cultivation/vegetation in each sample site.

Which site do you anticipate will have the most acidic pH? Why?

Which site do you think will have the most alkaline pH? Why?

Procedure for Measuring Soil-water pH (Cahoon, 1974)

1. Weigh out 10 g of air-dried soil from Sample A.
2. Rinse a 25-ml beaker with distilled water and dry.
3. Place your 10 g soil sample in the beaker and add 10 ml of distilled water to the soil. (*Note:* If the soil has a high clay content, you may need to let it settle overnight.)

Sample/site	Sample A cultivated field with fertilizer	Sample B forest soil	Sample C undisturbed fencerow or pasture
Sample pH			

Table 4-1. Results of soil-water pH for three different soil samples

4. Mix your soil and water in the beaker for 5 seconds.
5. Let the mixture stand for 10 minutes.
6. Gently swirl the soil-water mixture and lower the pH electrode so that it is just above the soil particles. If you are using pH paper, be certain that your paper is at this point in the beaker.
7. Allow the pH meter to come to equilibrium and take a reading.
8. Rinse the pH electrode with distilled water.
9. Repeat the procedure with Samples B and C and record the results in Table 4-1.

Did your results support your hypothesis?

Based on the results in Table 4-1 and what you know about soils, how do you think soil pH is related to vegetation type and/or agricultural practices?

DETERMINING SOIL TYPE AS A FUNCTION OF SOIL TEXTURE

As we know, there are several different particle-size classes: gravel, sand, silt, and clay. These particle-size classes are based on the diameter of the particle. The relative proportion of particle sizes in soil gives the soil many of its physical characteristics such as texture, pore size, water infiltration rate, water-holding capacity, and **cation-exchange capacity**, for example. Using a scale and a set of sieves with different pore sizes, soil type (as a function of soil texture classes—see Figure 4-1) can be determined. Your teacher will give you three soil samples: Sample A—cultivated field with nitrogen fertilizer added recently; Sample B—forest soil; and Sample C—soil from an undisturbed fencerow or pasture. These are the same sites that were used for soil pH.

Hypothesis

Write a hypothesis that relates the soil texture to the type of vegetation/agricultural practice used.

Which sample do you expect to have the highest proportion of clay? Why?

Which sample do you expect to have the highest proportion of sand? Why?

Procedure for Determining Soil Relative to Particle Size

1. Obtain a set of sieves from your instructor with the following pore sizes: 2.00 mm, 0.5 mm, and 0.002 mm. Stack the sieves so that the 2.00 mm pore sieve is on top, the 0.05 mm sieve is next, and the 0.002 mm sieve is on the bottom. Place the stacked sieves in a pan.
2. Obtain four small paper bags from your teacher. Label the bags as follows: SAMPLE A, >2.00 mm; SAMPLE A, 2.00–0.5 mm; SAMPLE A, 0.5–0.002 mm; and SAMPLE A, <0.002 mm.
3. Weigh out 10 g of oven-dried soil from Sample A.
4. Place soil from Sample A in the stacked sieves. Gently shake the sample so that the soil falls through the sieve (some particles will be retained in each screen).
5. The particles in the top sieve are greater than 2.00 mm (gravel); pour these soil particles into the bag labeled SAMPLE A, >2.00 mm.
6. The particles in the next sieve are smaller than 2.00 mm in diameter and larger than 0.5 mm (sand). Pour these particles into the bag labeled SAMPLE A, 2.00–0.5 mm.

Weight of soil fraction (g)	Sample A—cultivated field with fertilizer	Sample B—forest soil	Sample C—undisturbed fencerow or pasture
Gravel (> 2.00 mm)			
Sand (2.00–0.5 mm)			
Silt (0.5–0.002 mm)			
Clay (0.002 mm)			

Table 4-2. Soil fraction weights for three different soil samples

7. The particles in the next sieve are smaller than 0.5 mm in diameter and larger than 0.002 mm (silt). Pour these particles into the bag labeled SAMPLE A, 0.5–0.002 mm.
8. The particles in the bottom of the pan are smaller than 0.002 mm in diameter and represent the clay fraction of the soil. Pour these particles into the bag labeled SAMPLE A, <0.002 mm.
9. Weigh the soil in each of your four paper bags and record their weights (g) in Table 4-2.
10. Repeat the procedures for soil from Samples B and C. Remember to use new paper bags each time to get the most accurate weight estimates. Record the weights (g) of each texture class in Table 4-2.

Did your results support your hypothesis?

Calculate the percentage of each size class for each sample.

$$\% = \frac{\text{Weight of Soil Fraction (g)}}{\text{Total Weight of Sample (10g)}}$$

See Figure 4-1, soil texture triangle. What type of soil is found in Sample A?

What type of soil is found in the area where Sample B was taken?

What kind of soil was found in the area from which Sample C was taken?

Given your understanding of the relationship between soil pH and cation-exchange capacity, do you think that the relative proportion of gravel, sand, silt, and clay in the soil has an effect on soil water pH? Relate your answer to the data from Table 4-1.

ESTIMATING WATER-HOLDING CAPACITY FOR DIFFERENT SOILS

Water is essential for all plant growth. Animals also require water for survival. Many animals rely on stored water from plants as their primary water supply. How does the type of soil relate to the amount of water the soil can hold? Is it possible for soil to hold too much water? Soil is made of structural units called **peds**, which are described by their shape: aggregate, crumb, prism, block, plate, or granule. This is in contrast to soil texture, which is related to particle size. Factors that determine the structure of soils are soil texture (relative proportion of different particle sizes), organic matter content, amount of calcium carbonate and/or iron oxides, freeze-thaw cycles, biological activity (impact of earthworms, fungi, and bacteria in soil), and soil cultivation practices. Soils have spaces between the peds through which water, water-soluble minerals and organic nutrients, and atmospheric gases can move. When these spaces are completely filled with water, the soil is saturated. Eventually, atmospheric gases (oxygen—O_2, carbon dioxide—CO_2, and nitrogen—N_2) replace some of this water. Most plants do not thrive in poor drainage conditions resulting in permanently saturated soils. Waterlogged soils typically are low in O_2. Only plants and microorganisms adapted to low oxygen conditions survive in such locations. Swamps, bogs, and tidal marshes are examples of areas with permanently saturated soils under normal conditions. These areas do not make good sites for the cultivation of most plant crops. The amount of water that soils can hold is dependent upon the ped size, soil texture, and **slope** of the site. Sites with moderate to steep slopes drain water faster than flat areas. Using soil samples from three sites: a cultivated field, a forest, and a fencerow or pasture, you will estimate the amount of water in a saturated sample.

Sample site	Wet weight	Dry weight	Water (g) = wet weight – dry weight
Sample A—cultivated field with fertilizer			
Sample B—forest soil			
Sample C—undisturbed fencerow or pasture			

Table 4-3. Water saturation estimates for three different soil samples

Hypothesis

You have classified the soil types for each of these three sites based on soil texture.

Write a hypothesis regarding the relationship between soil type and the amount of water necessary to saturate the soil.

Which site do you think will require the most water for saturation? Why?

Procedure for Measuring Soil Saturation

1. Measure 100 ml of soil from each of the three sites: A, B, and C.
2. Place each sample in a canvas bag and tie the bag tightly and weigh the sample. Record the "dry" weight in Table 4-3.
3. Thoroughly saturate each bag with water by placing it underneath running water in the sink or putting the bag in a pan of water and allowing the bag to absorb as much as it can (approximately 20–30 minutes).

4. When the samples are saturated, allow the excess water to drain.
5. Weigh each of the samples again and record the "wet" weight in Table 4-3.
6. Calculate the weight of the water by subtracting the dry weight from the wet weight and record the amount in Table 4-3.

Did your results support your conclusion?

How do you think vegetation cover affects soil saturation?

SUMMARY

Given the measurements of soil-water pH, soil texture, and water saturation for each of these three sites, assuming that all other factors are the same (aspect, slope, climate), which of the three sites is best suited for cultivation? Include in your answer all of the evidence that you have gathered in the three experiments on physical soil characteristics.

Has the type of vegetation found in each area affected the physical properties of the soils? How?

In which area would you expect to find the greatest diversity of living organisms in the soil? Why?

What recommendations could you make for producers who are interested in maintaining healthy soil based on these experiments?

GLOSSARY OF TERMS

cation-exchange capacity: A measure of the total number of cations that can be exchanged between the soil solution and the surface of mineral particles (usually clay), roots, and humus; measured in milli-equivalents (meq) per 100 grams of soil at pH of 7 (or other specified pH); normal range is 2 to 50 meq/100 g of soil.

clay: The smallest particle size class in soil; less than 0.002 mm in diameter.

humus: Organic matter in soil which is an advanced stage of decay; characterized by a high nitrogen (N) content and a high cation-exchange capacity.

inorganic: A compound that lacks both the elements carbon (C) and hydrogen (H); example: sodium chloride → NaCl.

gravel: Mineral pieces which are greater than 2.00 mm in diameter.

loam: Ideal soil texture for cultivation, consists of less than 52% sand, 28% to 50% silt, and 7% to 27% clay.

organic: A compound that contains both the elements carbon (C) and hydrogen (H) and is derived from living matter; example: methane → CH_4.

ped: Unit of soil structure, peds are described by their shape: aggregate, crumb, prism, block, plate, or granule.

pH: A measure of the relative concentration of hydronium (H^+) ions. The greater the H^+ concentration, the more acidic a substance is; the lower the concentration of H^+ ions, the more alkaline a substance is.

1		7		14
very acidic	acidic	neutral	alkaline	very alkaline

sand: Mineral particles that are between 2.00 and 0.05 mm in diameter; may also be a textural class of soils that contains at least 85% sand and less than 15% clay particles.

silt: Very small mineral particles in the soil; USDA System—particles between 0.05 and 0.002 mm in diameter; International System—particles between 0.02 and 0.002 mm in diameter; also relates to the textural classification of soils—silty soil contains at least 80% silt and less than 20% clay particles.

slope: The angle or inclination of the land; calculated as the change of vertical distance with respect to change in horizontal distance; slope is expressed as a percentage:

$$\text{Slope} = \frac{\text{Vertical Distance}}{\text{Horizontal Distance}} \times 100\%$$

subsoil: The portion of the soil that lies below cultivation; the portion of the soil profile in which a change occurs in the texture and appearance of the soil.

topsoil: Surface and subsurface soils rich in organic material; fertile portion of the soil profile; this part of the soil is used for crop production.

EXERCISE 5

Soils: Biological Properties of Soil

OBJECTIVES

Upon completion of this lab exercise, you should be able to:

- identify the portion of the soil horizon containing the greatest concentration of microorganisms.
- define the role of algae, bacteria, and fungi in soils.
- distinguish between algal, bacterial, and fungal cells using a compound light microscope.
- describe the impact of agricultural practices on soil microbes.
- observe *Rhizobium* in the root nodules of a legume.
- estimate population numbers from a given soil sample using serial dilution microbial culture techniques.
- describe the role of soil macroorganisms such as earthworms, nematodes, and anthropods.
- observe soil macroorganisms collected in pitfall traps.

Suggested Reading:

Chapter 2 of *The Science of Agriculture: A Biological Approach*, 4th edition

INTRODUCTION

The organic material **humus** is found near the top of the soil surface constituting about 3% to 8% of the topsoil. In a soil profile, it is the **A horizon** which contains the organic portion of the soil and the living soil organisms. All soils contain some organic material that is derived from living organisms. The soil organisms play an important role in recycling organic material. Larger soil organisms, such as earthworms and arthropods, are responsible for reducing the size of dead plant and animal material. Earthworms feed on fungus and bacteria growing on the surface of dead leaves; by eating the leaves, the earthworms reduce the physical size of the **detritus**. The decrease in size makes the leaf material more attractive for colonization by microorganisms such as bacteria and fungi. The bacteria and fungi chemically break down large organic molecules into their smaller subunits. It is the smaller molecules that are absorbed by plant roots.

Some bacteria associated with the roots of leguminous plants and certain cyanobacteria (blue-green algae) are also responsible for increasing soil nitrogen content through a process called **nitrogen fixation** (see Figure 5-1). Nitrogen is essential for plant growth; the proteins and nucleic acids (DNA and RNA) necessary for normal cell functioning contain nitrogen. Following nitrogen fixation, nitrogen is present in the soil in the form of ammonia (NH_3). Ammonia is toxic to most plants and animals. Some soil bacteria convert the toxic ammonia to nitrate (NO^-_3), the form of nitrogen that is best utilized by most plants. Many soil bacteria are also involved in the process of organic decomposition, which releases organic and inorganic nutrients from decaying material into the soil.

Some soil bacteria and fungi also establish special relationships with plants. For instance, many types of leguminous plants produce higher

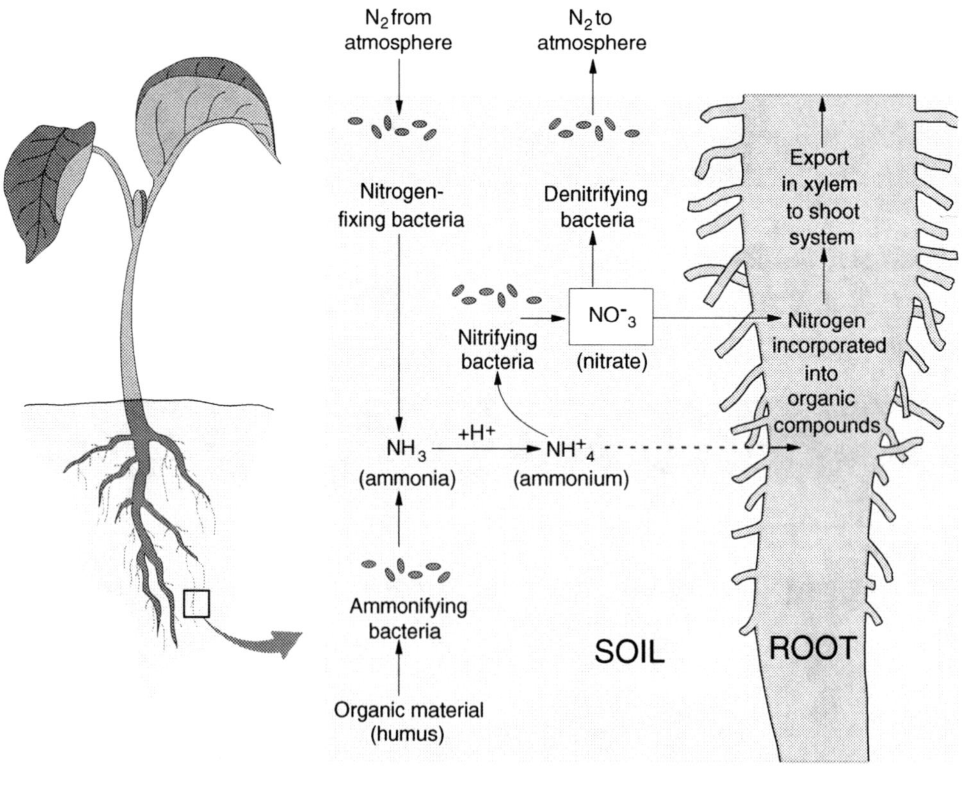

Soil bacteria form ammonium by fixing atmospheric N_2 (nitrogen-fixing bacteria) and by decomposing organic material (ammonifying bacteria). Although plants absorb some ammonium from the soil, they absorb mainly nitrate, which is produced from ammonium by nitrifying bacteria. The form of nitrogen transported up to the shoot system depends on the plant species. Most often, nitrogen from nitrate is incorporated into organic compounds such as amino acids in roots. This organic nitrogen is then exported to the shoot system via xylem.

Figure 5-1. The role of soil bacteria in the nitrogen nutrition of plants.

yields when bacteria are present in their roots (refer to Figure 5-2). Some fungi form **mycorrhizal** associations with plants, assisting the plants in the uptake of water and nutrients. Mycorrhizal fungi improve plant growth, and there are many examples of economically important mycorrhizal fungi, truffles for example.

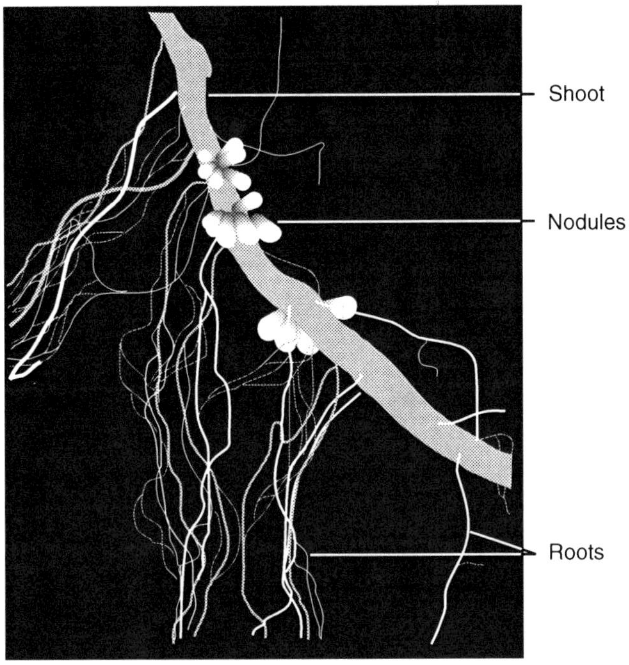

(**a**) The nodules of this pea root contain symbiotic bacteria that fix nitrogen and obtain photosynthetic products supplied from the plant.

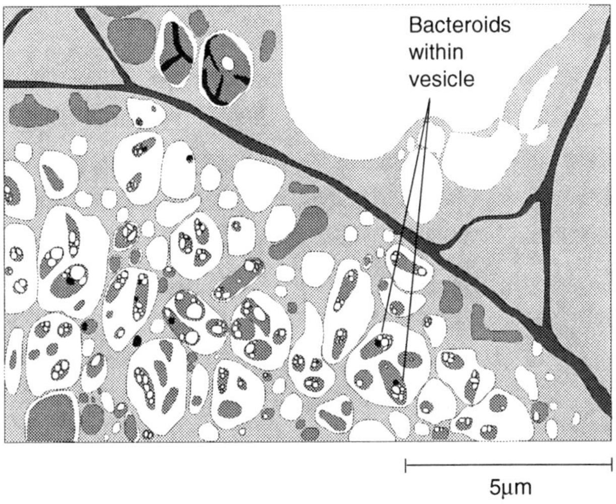

(**b**) In this electron micrograph, a cell from a root nodule of soybean is filled with bacteroids (TEM). The adjoining cell remains uninfected.

Figure 5-2. Root nodules of a legume, (a) labeled photograph and (b) TEM.

OBSERVING SOIL MICROORGANISMS

The presence of soil microorganisms provides an indication of soil quality. Healthy soils will have an abundance of soil bacteria, fungi, and algae. You will receive a soil-sterile water suspension from your instructor, which you will examine under the light microscope for the presence of soil microorganisms. Your teacher will provide you with three soil samples: Sample A—field under cultivation with leguminous plants, Sample B—mixed forest soil, and Sample C—soil from a pasture currently in use.

Hypothesis

Write a hypothesis that relates the kinds of soil organisms, bacteria, fungi, and algae to the type of cultivation/vegetation in each sample site.

1. With the cap secured, shake the soil-sterile water suspension from Sample A. Allow the large particles to fall to the bottom of the jar.
2. Using a sterile pipette, pipette a few drops of the soil-water suspension onto the middle of a flat slide.
3. With the cover slip at an angle, drop the cover slip on the soil-water suspension. (*Hint:* See Figure 3-5 if you are having difficulty.)
4. If you have air bubbles under your cover slip, gently tap the surface of the cover slip with your pencil eraser or the blunt end of a dissecting needle. The bubbles should move to the outside of the cover slip.
5. Turn on your microscope light and make any necessary adjustments (center the slide on the stage, adjust the light using the diaphragm, etc.). If you are having difficulty with your microscope, you might want to review the use of a compound light microscope in Exercise 3.
6. Begin with the low-power objective and bring the slide into focus. Bacterial cells are too small to be seen with the low-power objective, but fungi and some algae may be visible.
7. Without moving the slide, switch to the medium and high-power objectives. Remember: **Do not use the coarse adjustment knob with the high-power objective!**

8. Algal cells will appear rectangular or circular with a definite outside wall. The cells of soil algae are usually green, blue-green (very small-cell, blue-green algae also called cyanobacteria), or a brownish color (diatoms).

   Do you see any algae present in the soil-water suspension from Sample A? If so, draw what you see in the space provided.

9. Fungal cells are clear in color, appearing as long thin threads, **hyphae**, with definite walls between the cells. Some of the hyphae may have broken when you shook your sample jar.

   Do you see any fungal cells in the soil-water suspension from Sample A? If so, draw what you see in the space provided.

10. Bacterial cells are very small (150 to 500 μm). Many are too small for inexperienced eyes to see without the use of the oil immersion objective lens. Your teacher will review the use of the oil immersion objective lens with you. Be certain that the slide is in focus using the high-power objective lens, and adjust your condenser so that it is closest to the stage. Increase the amount of light by opening the condenser diaphragm. Move the revolving nosepiece in the direction of the oil immersion lens so that the slide is halfway between the high-power objective and the oil-immersion objective—DO NOT put the oil immersion objective in place yet. Put 1 to 2 drops of immersion oil on your cover slip and slide the oil immersion lens in place. Wait a minute or two to assure that the oil has covered the oil immersion objective. Using the fine adjustment knob only, bring the slide into focus. Bacterial cells come in three shapes (see Figure 5-3): spherical **(coccus)**, rod-shaped **(bacillus)**, and spiral **(spirilla)**. You should see single cells with a pronounced cell wall and little color.

    Do you see bacterial cells in your soil-water suspension from Sample A? If you see any bacteria cells, label the cells according to shape.

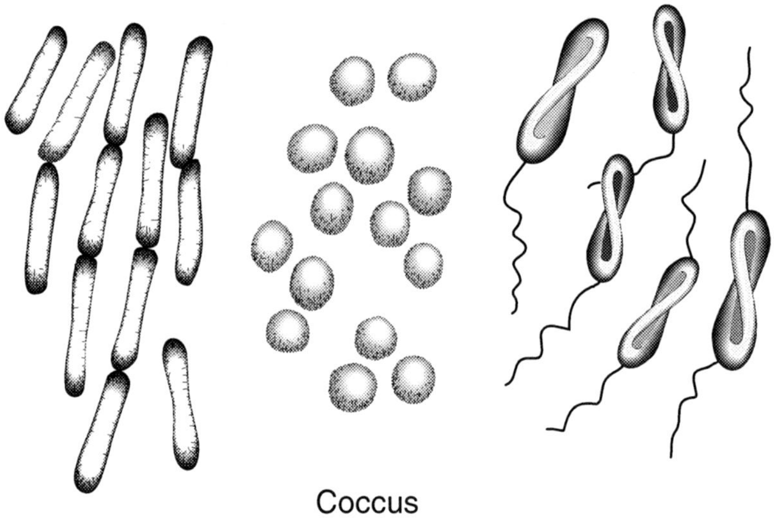

Figure 5-3. Shapes of bacterial cells.

11. Bacterial cells are easier to see with the aid of special colorizing agents called stains. Move the oil immersion lens out of place carefully and clean the immersion oil off the objective lens using the cleaning solution, provided by your teacher, and lens paper. Remove your cover slip and put 1 drop of crystal violet stain under your cover slip and add more water if necessary. Put a new, clean cover slip on your slide and focus your slide using the low, medium, and high-power objectives. Use the oil immersion objective as before, and look at your slide.

    Are there more bacterial cells visible with the crystal violet stain? How many different shapes of cells do you see?

12. Repeat steps 1 through 12 for Samples B and C.

    Were algae present in any of your soil samples? From which site?

What would the presence of algae suggest about the amount of water the soil is holding?

Were fungal hyphae observable in any of your soil samples? From which sample?

What would fungi indicate about the type(s) of vegetation in the area?

What roles do fungi play in soils?

Were bacteria present in any of your soil-water suspensions? In which sample(s)?

In which sample were the bacteria most prevalent?

In terms of your original hypothesis, how are vegetation cover and tillage practices related to the numbers and kinds of soil microorganisms?

OBSERVING LEGUME ROOT NODULES

Small amounts of available nitrogen in the soil can limit plant growth. Many important cash crops such as wheat, corn/maize, sorghum, cotton, tobacco, and oats have high nitrogen requirements. The practice of crop rotation, planting different crops in a field (or using the field as a pasture) from one growing season to the next, can greatly improve soil quality and reduce the amount of chemical fertilizer required for profitable yields. Practitioners of crop rotation frequently use leguminous plants as an alternative crop to winter wheat, corn, and cotton. Nitrogen-fixing bacteria belonging to the genus **Rhizobium** are found in association with bumps (nodules) on the roots of peas, beans, clovers, and other leguminous plants.

Hypothesis

In which of the three soil samples (Sample A—field under cultivation with leguminous plants, Sample B—mixed forest soil, and Sample C—soil from a pasture currently in use) would you expect available nitrogen to be greatest?

Explain how the amount of available soil nitrogen is related to the differences in cultivation practices among the three sites.

1. Obtain a **legume** root with visible nodules from your teacher. Using a dissecting needle, put 1 to 2 nodules on a clean microscope slide.
2. Put several drops of water and 1 drop of crystal violet stain on the nodules.
3. Add a cover slip. Wrap the slide in a paper towel—cover slip side up.
4. Using the eraser end of your pencil or the blunt end of your dissecting needle, crush the nodules until the cover slip is flat. Be careful not to break the cover slip.
5. Using the low-, medium-, and high-power objectives focus the slide.

6. Use procedure for the oil immersion objective lens, and look at the nodules at 1000× magnification.

   Draw what you see in the space provided. What shape are the *Rhizobium* bacteria (coccus, bacillus, or spirillum)?

ESTIMATING SOIL BACTERIAL POPULATIONS

As previously noted, the numbers and kinds of soil bacteria provide agricultural scientists with an indication of soil health and potential productivity. Techniques for estimating soil bacterial populations are useful tools in evaluating the suitability of soils for different agricultural practices. Your class will be divided into smaller groups, each group receiving a soil sample from one of the three sites: Sample A—field under cultivation with leguminous plants, Sample B—mixed forest soil, and Sample C—soil from a pasture currently in use. Using a microbiological lab technique called **serial dilution**, your group will estimate the number of bacteria present in 1 gram of soil from your site. You will compare the estimates of bacterial population size with the other groups in your class.

Hypothesis

Write a hypothesis that relates the size of soil bacterial populations to the type of cultivation/vegetation in each site.

1. Weigh out 1 gram of soil from your sample.
2. Remove the stopper from a sterile flask, add 1 gram of soil to 10 ml of sterile distilled water. Remember that you do not want to touch the bottom of the stopper or lay it on your table or lab bench; you want to avoid potential contamination of your soil-sterile water solution. Put the stopper back in place and shake the flask so that the soil is suspended throughout the water column.
3. Allow the large particles to settle to the bottom of the flask.

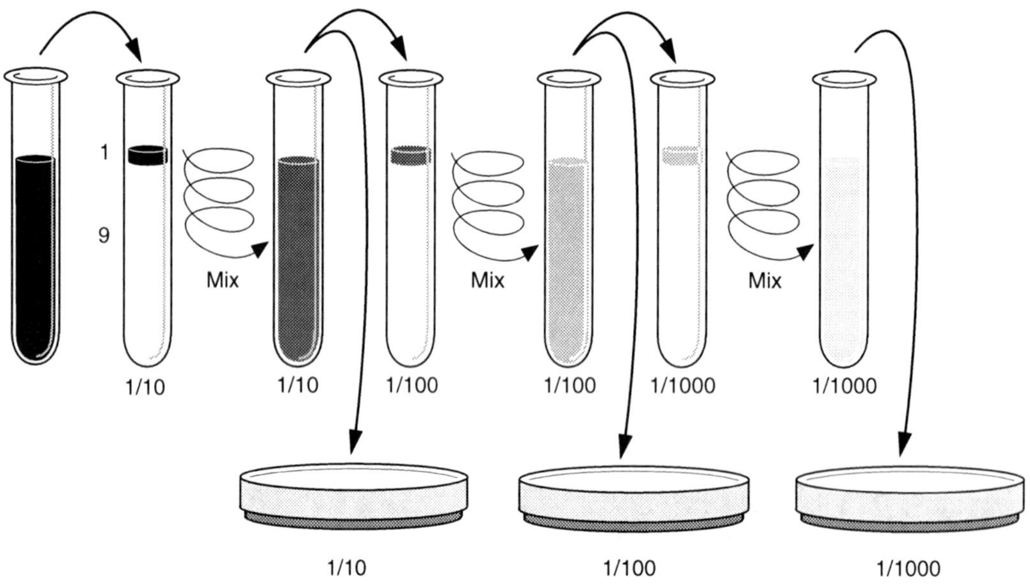

Figure 5-4. Serial dilution technique.

4. Take three test tubes, each containing 9 ml of sterile water and label them: 1/10, 1/100, and 1/1,000. Make a serial dilution of your original water sample as shown in the diagram. Use the same pipette for all three transfers but rinse it between each transfer; in other words, after discharging the contents of the pipette into the new dilution tube, rinse the pipette by drawing up some of the medium and emptying the pipette again. Remember to mix each tube before continuing the dilution.
5. Label six sterile petri dishes (two each) 1/10, 1/100, and 1/1,000 and add *your initials, lab period,* and *type of water used.*
6. Using a clean pipette and beginning with the 1/1,000 dilution, inoculate each of the six plates with 1 ml from the appropriate dilution tube.
7. The agar needs to be slightly above room temperature and liquid enough to pour. If the media is too hot, the microorganisms will be killed when the agar is added. Using aseptic technique, pour enough agar into one of the inoculated dishes to cover 2/3 of the plate. Immediately mix the molten agar with the diluted water sample by gently rotating the plate clockwise and counterclockwise without splashing or lifting it from the lab bench. Repeat the process with the other inoculated petri dishes. Allow the plates to cool until the agar is well hardened. Place a piece of tape on either side of each dish to prevent them from coming open. Each person is to take home three plates (one plate of each dilution) and maintain them at room temperature. After 48 hours count the colonies present on those plates where counting is feasible.

In the space provided, record the type of water used, the number of colonies present on each plate, and calculate the number of viable bacteria present per ml of the original soil-water sample, taking into account the predilution performed on the soil (10 ml of sterile water). Next week in lab, compare your numbers with those of your lab partner and the other students working on the same site.

Predilution, if any, for your sample (A, B, and C) 1 g soil/10 ml sterile water

Colonies present per plate:

1/10 _____ 1/100 _____ 1/1,000

Viable bacteria present per ml of the original soil sample:

_____ cells/1 g/ml

Collect the data from the class and record it in Table 5-1.

An average is calculated for each sample by adding all of the estimates for the site and dividing the total by the number of estimates. For example, if there were four groups with estimates of 150/ml, 250/ml, 500/ml, and 50/ml, the average for the site/sample would be [(150 + 250 + 500 + 50)/4]. You need to calculate the average population size for each of the three sample sites and record the values in the right-hand column in Table 5-1.

Was your hypothesis supported by the data the class collected?

Sample site	Bacterial population estimates	Average population size
Sample A cultivated field with leguminous plants		
Sample B mixed forest soil		
Sample C pasture currently in use		

Table 5-1. Bacterial population estimates from three different types of soils

Were your results different than expected? If so, give some reasons why the results may have been different than you initially predicted.

SOIL MACROORGANISMS

Arthropods are **invertebrates**, animals that lack a backbone. Arthropods have exoskeletons, that is, their skeleton is on the outside of the body. All arthropods also have jointed bodies and appendages. Insects, spiders, scorpions, ticks, and mites are all arthropods. Another common soil invertebrate is the nematode. Nematodes are a type of worm that lacks internal segmentation. In terms of numbers, nematodes are the most abundant invertebrates on earth. Nematodes can be found living independently in the soil, but most frequently, nematodes are found in association with plant roots. Some nematodes can cause extensive crop damage by injuring plant roots; this affects the plant's ability to take up water and nutrients. A nematode infestation can greatly reduce crop yield. Most soil invertebrates are beneficial; they incorporate the remains of plant material and humus through the uppermost portions of the soil profile.

There are many methods for collecting soil macroorganisms. One technique involves the use of core samples, long tubes of specific diameters sunk to a given soil depth. Core samples are used to look at soil **microflora**, microscopic invertebrates, bacteria, fungi, and algae. Macroscopic animals can be collected in the field using pitfall traps (see Figure 5-5). You will be given pitfall trap samples collected from the three sites you have been working with: Site A—a field under cultivation with leguminous plants, Site B—a mixed forest, and Site C—a pasture currently in use. You will use a dissecting microscope or a magnifying lens to categorize and count the invertebrates in your sample.

Hypothesis

Write a hypothesis that relates the number of different kinds of soil invertebrates to the type of cultivation practice in each of your three sites.

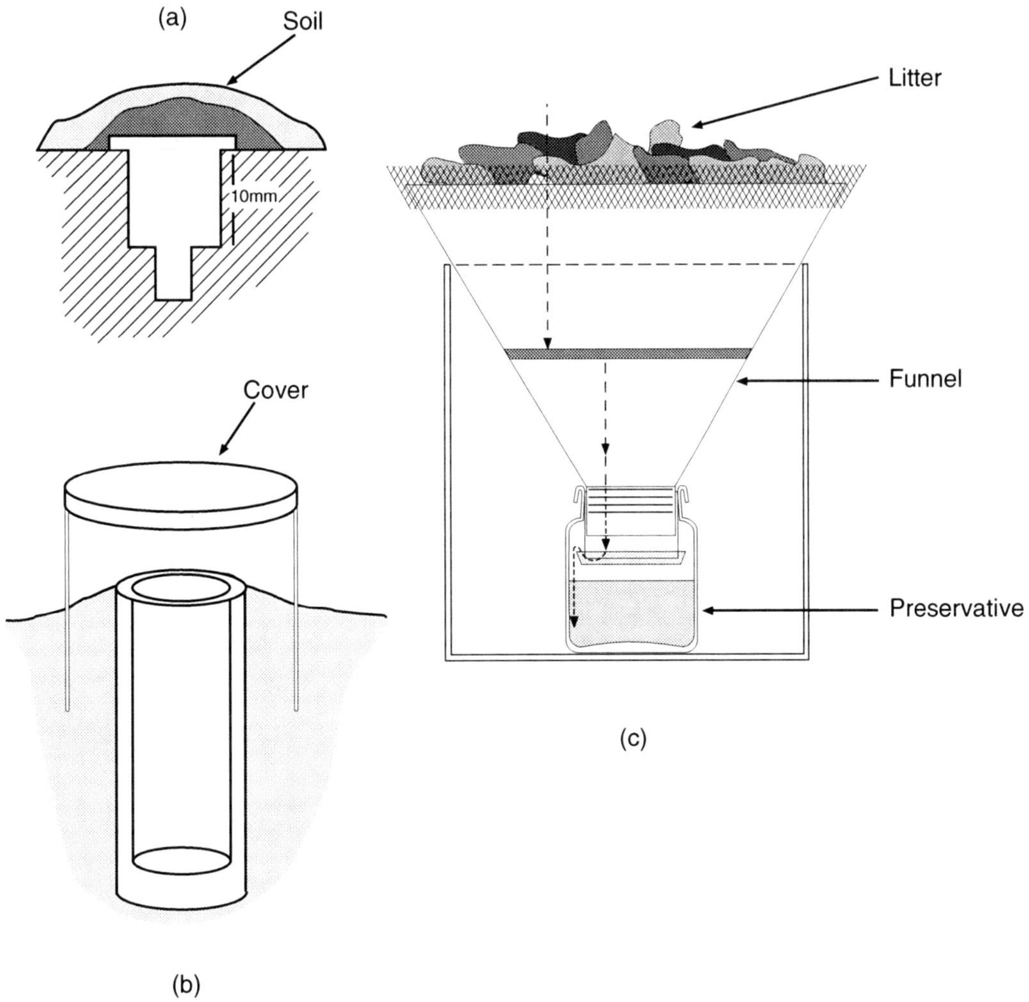

Figure 5-5. Kinds of soil pitfall traps, (a) Soil-covered trap, (b) standard pitfall trap, and (c) litter-covered funnel trap.

1. Empty the organisms from the trap bottle or vial (with alcohol used as a preservative) into the dish provided by your teacher.
2. Place the dish under the dissecting microscope (or alternately use your hand lens).
3. Separate your animals into groups that look different. Place all of the individual animals belonging to one kind of invertebrate group into smaller bottles with preservative from your teacher. Your teacher will help you to determine which animals are annelids (earthworms), nematodes, and arthropods (crustaceans [such as pill bugs]); millipedes, centipedes, arachnids (spiders, ticks, and mites), and insects. Record the number of different kinds of soil macroorganisms for each of your three sites in Table 5-2.

Did your observations support your original hypothesis?

Type of invertebrate	Site A: Field under cultivation with leguminous plants	Site B: Mixed forest	Site C: Pasture currently in use
Nematodes			
Annelids			
Arthropods: Crustaceans			
Arthropods: Millipedes			
Arthropods: Centipedes			
Arthropods: Arachnids			
Arthropods: Insects			

Table 5-2. Numbers of different types of soil invertebrates from three different sites

How do you think the presence of soil microorganisms influences the numbers and kinds of soil macroorganisms?

GLOSSARY OF TERMS

A horizon: The uppermost mineral layer of the soil profile; contains the topsoil organic material mixed with the minerals derived from the parent material.

bacillus: (pl. *bacilli*) A rod-shaped bacterium.

coccus: (pl. *cocci*) A spherical-shaped bacterium.

detritus: Loose fragments of material which results from the decomposition of organic matter.

humus: Organic matter in soil which is in an advanced stage of decay; characterized by a high nitrogen (N) content and a high cation-exchange capacity.

hypha: (pl. *hyphae*) (*hyphos*—web, to weave) The multicellular threadlike filaments that comprise the body (mycelium) of a fungus.

invertebrate: (*in*—without or lacking; *vertebratus*—spinal column, vertebrae) Animals that do not have a spinal cord enclosed in a bony column; sponges, worms, nematodes, insects, snails, starfish, and squid are all examples of invertebrate animals.

legume: (*legere*—to gather) The fruit or seed of a plant in the pea family. Leguminous plants include many agriculturally important plants: clover, peanuts, soybeans, peas, beans, alfalfa, lespedeza, and vetch. These plants aid in the return of organic nitrogen to the soil as their roots contain nitrogen-fixing bacteria.

mycorrhizae: (*myco*—fungus; *rhiza*—root) A symbiotic relationship between a fungus and plant roots, usually a tree or shrub. Mycorrhizal relationships are more common in areas where the soil is poor in nutrients. The presence of the fungus allows the plant to take advantage of the reduced nutrient concentration in the soil.

nitrogen fixation: The conversion of atmospheric nitrogen (N_2) to ammonia (NH_3) by certain bacteria associated with the roots of leguminous plants and cyanobacteria (blue-green algae) with heterocysts.

Rhizobium: (*rhiza*—root) A group of small heterotrophic soil bacteria which are capable of invading the roots of leguminous plants, forming nodules on the roots. Once incorporated in the root, these bacteroides fix atmospheric nitrogen.

serial dilution: A technique in which a series of samples, each a known fraction of the previous sample, are diluted using sterile water or culture media. Usually, the dilutions related to a starting concentration/volume are fractions of 10; 1 ml, 1/10 ml, 1/100 ml, 1/1,000 ml for example.

spirillum: (pl. *spirilla*) A corkscrew or spiral-shaped bacterium.

EXERCISE 6

Cells: Prokaryotic Diversity

OBJECTIVES

Upon completion of this lab exercise, you should be able to:

- distinguish between prokaryotic and eukaryotic cells.
- recognize the three morphological shapes of bacterial cells.
- distinguish between Gram-positive and Gram-negative bacteria.
- identify the antibiotic which is most effective against a Gram-positive and Gram-negative bacterium.
- draw a cyanobacterium (blue-green alga).
- describe the function of akinetes, heterocysts, and endospores.
- describe the symbiotic relationship between the water fern, *Azolla*, and the cyanobacterium, *Anabaena azollae*, and explain the agricultural implications of the *Azolla-Anabaena* relationship.

Suggested Reading:
Chapter 3 of *The Science of Agriculture: A Biological Approach*, 4th edition.

INTRODUCTION

The building blocks of living organisms are cells. Cells are the smallest unit of life capable of independent existence. The smallest of all cells belong to the bacteria. Bacterial cells are simple and lack most of the internal compartments, **organelles**, present in other cells. Bacterial cells do not have a central nucleus enclosed within a membrane. The bacterial chromosome is found within the cytoplasm of the cell. Bacteria also lack other membrane-bound organelles common to most cells. Bacteria and cyanobacteria (blue-green algae) have **prokaryotic** cells. In contrast, all other living organisms have cells with a distinct membrane-bound nucleus and specialized compartments surrounded by membranes or organelles. The cells of protists, fungi, plants, and animals are **eukaryotic** cells. In this exercise, you investigate the properties of prokaryotic cells. Because these cells are very small, you will need to use the compound light microscope to make your observations.

Bacterial cells are classified according to shape. Spherical cells are **cocci**, rod-shaped cells are **bacilli**, and corkscrew-shaped cells are **spirilla**. Microbiologists who study bacteria give bacteria names that describe the shape of the colonies. Strepto- refers to bacterial cells that form a long chain, diplo- describes bacterial colonies in which the cells appear in groups of two, and staphylo- is used when the bacterial cells form aggregates. Using this logic, diplococcus would be the name used to describe a bacterium that has spherical cells in groups of two. Using the prefixes strepto-, diplo-, and staphylo-, and the three bacterial shapes, fill in the table provided with the appropriate name.

Description of bacterial cell shape	Name
Chain of rod-shaped cells	
Two rod-shaped cells	
Cluster of round cells	
Corkscrew-shaped cells	
Chain of round cells	

Microbiologists also use staining techniques to help observe and identify bacteria. One technique is called the Gram stain. The Gram stain divides bacteria into two groups: Gram-positive and Gram-negative bacteria (see Table 6-1). The bacteria are placed on a slide and

Characteristic	Gram-positive	Gram-negative
Color	Violet to black	Pink to red
Peptidoglycan content	90% of cell wall	5–20% of cell wall
Sensitivity	to penicillin	to streptomycin
Shape	Spore-forming rods, many cocci	Most nonspore-forming rods, spirals, some cocci
Toxins	Exotoxins (if any)	Endotoxins
Susceptibility to lysozyme	Very susceptible	Less susceptible

Table 6-1. Properties of Gram-positive and Gram-negative bacteria. These properties result directly from the structure and chemical constituents of the bacterial cell wall.

Bacillus (rod)		Coccus (round)		Spirillum (spiral)
Gram-positive	**Gram-negative**	**Gram-positive**	**Gram-negative**	**Gram-negative**
Bacillus cereus	*Escherichia coli*	*Sarcina lutea*	*Neisseria perflava*	*Rhodospirillum rubrum*
Bacillus megaterium	*Pseudomonas aeruginosa*	*Staphylococcus aureus*		
Bacillus subtilis		*Staphylococcus epidermidis*		
		Streptococcus faecalis		

Table 6-2. Gram stain reaction for some common species of bacteria.

a small amount of primary stain, crystal violet, is added. Excess stain is removed with a few drops of water. A few drops of Gram's iodine are added to fix the stain. Adding a few drops of an alcohol-acetone mixture then destains the slide. A second stain, safranin (red), is applied to the slide. Bacteria whose cell walls bind the crystal violet stain remain purple in color; these are **Gram-positive bacteria**. Bacteria with a different kind of cell wall lose the purple stain and appear pink—these are **Gram-negative bacteria**.

OBSERVING BACTERIA

1. Obtain a prepared slide of a mixed bacterial culture which has been Gram stained.
2. Using the compound light microscope, focus on the slide first using the 10× (4× if your microscope has one).
3. Follow the procedures outlined by your teacher for the use of the oil-immersion objective lens (100×). Using the oil-immersion lens, focus the slide until pink and purple cells become visible.

Are there more Gram-positive (purple) or Gram-negative (red) cells on your slide?

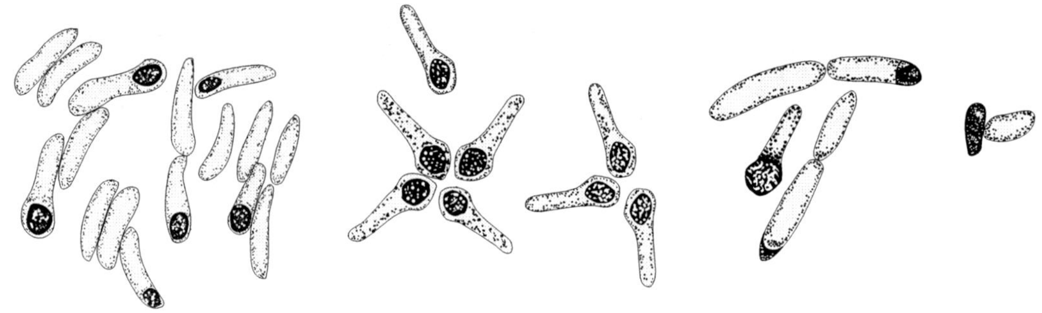

Figure 6-1. Bacteria with endospores.

Draw and label several clusters of bacterial cells using the descriptive terms: bacillus, coccus, and spirilla. If the bacterial cells appear in groupings, also include the appropriate descriptive prefix (diplo-, strepto-, or staphylo-).

Antibiotics are substances that inhibit the growth of microorganisms. Most antibiotics are used as medicines for commercially important plants and animals, or for humans. Many antibiotics work by changing the cell walls of bacteria in such a way that the bacterial cells lose their ability to divide. If the Gram-staining technique is based on differences in bacterial cell wall composition, how might physicians and veterinarians use this technique when deciding which antibiotics to use to treat a patient? (*Hint:* Look at Table 6-1.)

Some bacteria have the ability to produce **endospores**, small asexual reproductive structures inside of the cell. Endospores are a response to environmental stress; the endospore can survive unfavorable conditions much longer than the bacterial cell from which it was derived. Obtain a prepared slide of bacteria with endospores, which has been stained so that the endospores are visible. Remember that the endospores are inside the bacterial cell (see Figure 6-1); an endospore is extremely small, and it is visible only with the oil-immersion lens. Draw and label a bacterial cell with an endospore in the space provided.

ANTIBIOTIC SENSITIVITY TESTS

Week 1: Set Up Experiments

Place a small paper disk treated with each test compound in one quadrant of a plate of nutrient agar that contains the bacterium. If the compound is effective, a clear ring of no growth, the **zone of inhibition**, appears around the disk after the bacterial plate is incubated (see Figure 6-2).

Use **aseptic technique** at all times when handling these plates! Follow directions carefully.

1. You and a partner will receive two petri dishes of nutrient agar (*Note:* Black marks on the sides of the petri dishes).
2. Write your names and lab period on *lids* (the larger dish) of all three petri dishes.

   Number the dishes 1 and 2.
3. Turn over dishes 1 and 2. Use a grease pencil to draw a quadrant on the outside of each agar-containing (smaller) plate. (Do *not* place quadrant on the lid. Why?)
4. Inoculate the agar of plate 1 with the Gram-positive bacteria available. Use aseptic technique.

   Label which bacteria you use on the bottom of the agar-containing plate.
5. Inoculate the agar of plate 2 with the Gram-negative bacteria available. Use aseptic technique.

   Label which bacteria you use on the bottom of the agar-containing plate.

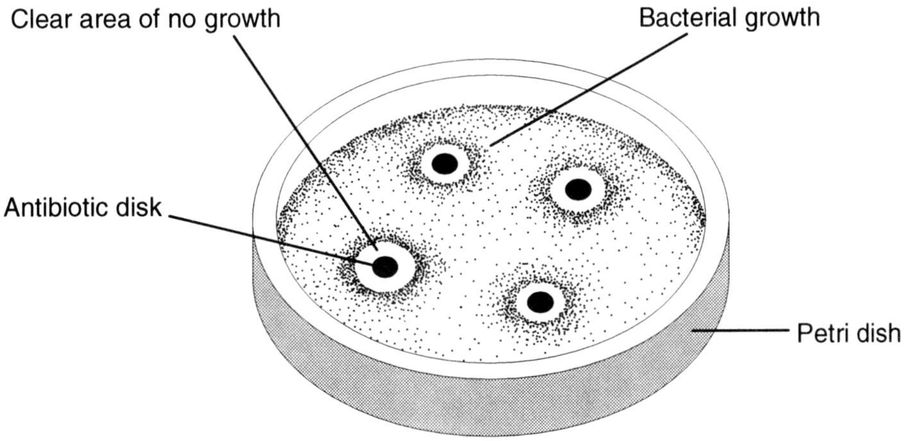

Figure 6-2. Bacterial growth in a petri dish showing zones of inhibition.

6. Place one antibiotic disc in the center of each quadrant, using forceps to transfer the four different antibiotic discs. The letters on the discs identify which antibiotic they contain.
7. Be sure the discs adhere well to the agar, but do not push them into the agar.

Hypothesis

You are using a Gram-negative bacteria, *E. coli,* and a Gram-positive bacteria, *B. subtilis,* to test the effectiveness of four antibiotics: penicillin, streptomycin, tetracycline, and ampicillin. Using the information contained in Tables 6-1 and 6-2, write a hypothesis about the effects of these antibiotics on the two different bacteria.

TEST OF ANTIBIOTIC AND ANTISEPTIC ACTION

Week 2: Analyze Results of Experiments

1. Measure the diameter of the zone of inhibition of growth of the bacteria for each antibiotic and antiseptic you tested.
2. Record this diameter in the table provided. You are responsible for the other class data. You will be asked to average your data with the numbers collected by the other members of your class.

   In the column to the left under zone of inhibition, record your results. The right-hand column will be used to record class data for each of the four antibiotics tested. Sizes of zones of inhibition indicate the relative effectiveness of different antibiotics and antiseptics against bacteria. The largest zone is the most effective antibiotic against the bacteria.

Antibiotics	Zone of Inhibition Diameter for *E. coli* (mm)		Zone of Inhibition Diameter for *B. subtilis* (mm)	
	Individual data	Class data	Individual data	Class data
Tetracycline				
Streptomycin				
Penicillin				
Ampicillin				

Which antibiotic was most effective against *E. coli?*

Which antibiotic was most effective against *B. subtilis?*

Is there an antibiotic that is effective against both bacteria?

With respect to your hypothesis about the effects of penicillin, streptomycin, tetracycline, and ampicillin on these two bacteria, was your original hypothesis supported by your class results? If not, explain why.

CYANOBACTERIA

Cyanobacteria (also called blue-green algae) are prokaryotic cells that make their own food by the process of photosynthesis. In addition, some species of blue-green algae also fix atmospheric nitrogen into an organic form usable by plants. Nitrogen-fixing cyanobacteria have specialized cells, **heterocysts**, which contain the enzymes responsible for the conversion of atmospheric nitrogen to ammonia (see Figure 6-3). Cyanobacteria are found in the soil and water; some species can be responsible for the "blooms" of algae, which cover pond surfaces and give the water an "off" odor and taste. A few species of *Anabaena* and *Nostoc* produce substances that are toxic to livestock and humans.

In Asia and India, a water fern, Azolla, is grown in conjunction with rice in flooded fields. Inside the floating leaves of the Azolla is the cynobacterium *Anabaena azollae*. The *A. azollae* is an active nitrogen-fixer, providing the water fern with a nitrogen source. The flooded tropical soils are low in nitrogen, and this tends to limit the productivity of the fields. Researchers in China, the Philippines, Vietnam, and India have shown increased rice yields when *Azolla* is used as a cover crop or in rotation with rice (Roger et al., 1993). The *Azolla* is used as a green fertilizer, supplying the rice with an organic nitrogen source (from the *Anabaena)* as it decomposes. The relationship between the water fern and *Anabaena* is an example of **symbiosis** (mutualism). In this relationship, both the *Anabaena* and the *Azolla* benefit.

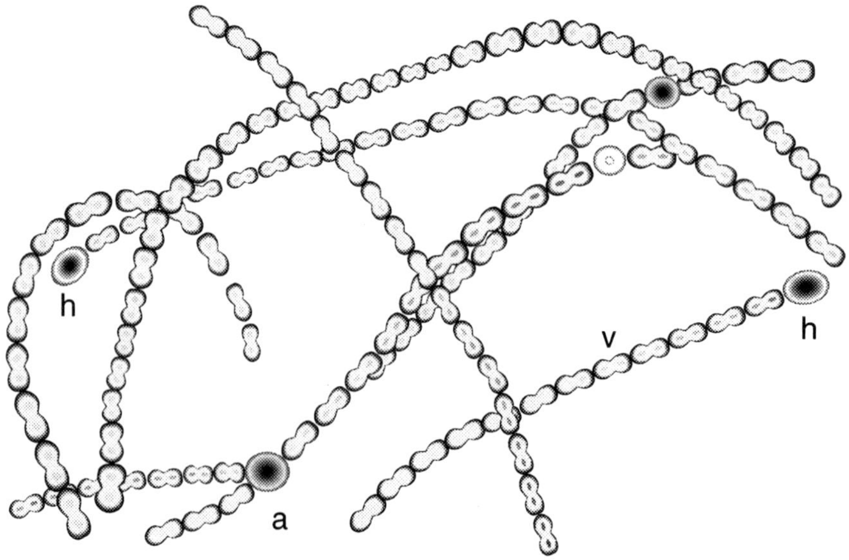

Figure 6-3. *Anabaena azollae*, filament with heterocysts (h), vegetative ceils (v) and Akinete (a).

OBSERVING *AZOLLA-ANABAENA* SYMBIOSIS

1. Put several *Azolla* specimens on a slide.
2. Place a second slide on top of the *Azolla* and gently press the two slides together, pulverizing the *Azolla*.
3. Separate the two slides and add a drop of fresh water to the crushed *Azolla*.
4. Place a cover slip on the specimen. Make observations of your slide using the 10× and 40× objectives. Draw and label what you see in the space provided (see Figure 6-3).

Representative Genera of Cyanobacteria

Prepare a wet mount of the mixed suspension containing several different genera of cyanobacteria. Based on the following descriptions, use the space provided to draw and label as many genera as you can find.

Gloeocapsa-unicellular, products of recent divisions often contained within a common sheath.

Oscittatoria-unbranched filament of closely appressed, disk-shaped cells; may exhibit gliding motility in which the tip appears to oscillate; **akinetes** and heterocysts absent.

Anabaena-unbranched filament of round cells (resembling beads on a string); akinetes and heterocysts may be present.

Fischeretta-branched filaments; akinetes and heterocysts may be present.

Gloeotrichia-tapered, unbranched filament with heterocyst at wide end of filament.

Also estimate and record the length of an *Andbaena* vegetative cell

____________________. (Don't forget to include the units.)

Hint: Look at Exercise 3 for the diameter of the field of view for each objective lens. Estimate the fraction of the field of view that a single vegetative cell occupies and multiply that fraction times the field diameter at that magnification.

Based on your observations, suggest practical agricultural uses for cyanobacteria and bacteria.

GLOSSARY OF TERMS

akinete: A thick-walled large cell found in cyanobacteria which contains a large food store; produced in response to unfavorable environmental conditions and allows the organism to survive until the environment is more favorable for growth.

antibiotic: (*anti*—against, *biotikos* pertaining to life) An organic molecule produced by a microorganism that inhibits or retards the growth of other microorganisms; may be produced synthetically from derivatives of microorganisms.

aseptic technique: (*a*—without, *sēptikos*—putrefying) Laboratory method for working with microorganisms. This method requires a meticulously clean technique; hands must be washed and gloved; lab bench surfaces must be cleaned prior to and after working with any living cultures; culture loops must be flamed before and after working with cultures; and the mouths of culture tubes must be flamed before the tube is inoculated and following inoculation. Your lab instructors will demonstrate aseptic technique for you prior to working with any of the living cultures.

bacillus: (pl. *bacilli)* A rod-shaped bacterium.

coccus: (pl. *cocci)* A spherical-shaped bacterium.

endospore: (*end*—Within*)* An asexual reproductive structure produced in some bacteria in response to environmental stress.

eukaryotic: (*eu*—true, *karyon*—kernel) A cell that contains a nucleus surrounded by a membrane and membrane-bound organelles (chloroplasts and mitochondria, for example); present in protists, fungi, plants, and animals.

Gram-negative: Bacterial cells that appear pink when the Gram-staining technique is used.

Gram-positive: Bacterial cells that stain purple when the Gram-staining technique is used.

heterocyst: (*hetero*—different) A specialized, thick-walled cell in some filamentous cyanobacteria which fixes nitrogen.

organelle: A specialized structure within a cell which performs a specific function; organelles are present in both prokaryotic and eukaryotic cells.

prokaryotic: (*pro*—first, *karyon*—kernel) A cell which lacks a membrane-bound nucleus or membrane-bound organelles; bacteria and cyanobacteria have prokaryotic cells.

spirillum: (pl. *spirilla)* A corkscrew- or spiral-shaped bacterium.

symbiosis: (*sym*—together, along, with; *bios*—life) A condition in which two or more organisms of different species live together in close contact.

zone of inhibition: Distance around an antibiotic (or other antimicrobial substance) disc placed in a microbial culture in which no microbial growth is observed.

EXERCISE 7

Eukaryotic Cells: Diversity in Structure and Function

OBJECTIVES

Upon completion of this lab exercise, you should be able to:

- discuss the taxonomic classification of living organisms including domain, kingdom, phylum, class, order, genus, and species.
- distinguish between prokaryotic and eukaryotic cells.
- give an example of a prokaryotic organism.
- give an example of an organism with eukaryotic cells.
- describe the structure and function of a eukaryotic cell nucleus.
- give the function of the following cellular organelles:
  - cell (plasma) membrane
  - chloroplast
  - cilia
  - flagella
  - mitochondria
  - vacuole (several types: food and contractile)
- differentiate between autotrophic and heterotrophic modes of nutrition.
- describe the typical protistan cell structure.
- describe the typical fungal cell structure.
- describe the typical plant cell structure.
- describe the typical animal cell structure.

Suggested Reading:
Chapter 3 of *The Science of Agriculture: A Biological Approach*, 4th edition.

INTRODUCTION

What cells do and how they do it has interested biologists since about 1670 when Anton van Leeuwenhoek first observed living cells with a microscope. Since then, we are convinced that all organisms consist of cells. We have also learned that *cellular form and function are intimately related.* Cells that do similar things look alike. Cells that do different things look different.

When viewing cells in a microscope, a cell biologist can identify whether they are bacterial, fungal, plant, animal, or protist; and often to what organ and tissue they belong and what they do. Pathologists routinely identify cancer cells in the microscope by looking for the features that indicate rapid growth. In this laboratory, you will learn some of the features of eukaryotic cells that identify them to biologists.

Fossil records indicate that the first bacterial cells appeared around 3.8 billion years ago. The earliest forms of life were all **prokaryotic** cells. Recent molecular studies suggest that all living organisms are derived from a universal prokaryotic ancestor, and three separate independent lines have been evolving for the past 1.5 billion years. These separate evolutionary lines are used to classify all living organisms into one of three **domains** (most general classification). Two domains, *Archaea* and *Bacteria,* are composed of prokaryotic organisms. In Exercise 6, you made observations about bacteria, and you will work with soil and milk microbes later in this course. Archaea are ancient bacteria living in extreme conditions such as geysers, hot springs, soda lakes, and in deep ocean vents. Scientists believe that modern-day archaea are found in environments most similar to those of ancient earth. The third domain, *Eukaryota,* contains all of the organisms that you will observe during this lesson: single-celled organisms called protists (Kingdom Protista), fungi (Kingdom Fungi), plants (Kingdom Plantae), and animals (Kingdom Animalia).

Eukaryotic cells characterize all members of the domain Eukaryota. Eukaryotic cells are organelles that perform different functions. Think of this diversity as a catalog of structures from which evolution and development has chosen. All cells possess a set of more or less standard **organelles** for keeping house. In virtually all eukaryotic cells, a **nucleus** contains genetic information in the form of DNA. A **cell membrane** (plasma membrane) governs what enters and leaves the cell, and it possesses components that help the cell communicate with neighboring cells. **Mitochondria** are organelles in which organic molecules are broken down to produce chemical energy for cellular processes.

The processes of evolution and development result in specialized cells by modification of these standard organelles. Plant cells that collect solar energy by photosynthesis use **chloroplasts** for this function. These organelles identify plant and algae cells. If you find cells with **cilia** or **flagella**, you can be sure either that these cells are motile or that the cilia functions to move substances around the outside of the cell.

Relatively little is known about how cells function and how they specialize. Research during the next 50 years will increase the understanding of this as new techniques are found that enable scientists to follow the instructions that genes send and receive from the cytoplasm. The payoff of such present research could be dramatic. Such knowledge could lead to prevention of cancer and AIDS. Universal organ transplants will probably become routine. Plants may be cultivated that grow human proteins. To understand what is coming, you must begin to learn how to deduce cell function from cell form in this lab.

OBSERVING LIVING CELLS USING A LIGHT MICROSCOPE

Now that you are familiar with some cellular structures common to both prokaryotic and eukaryotic cells, you are going to use some living specimens and the light microscope to make observations of several unicellular and colonial protists, and multicellular plant and animals. In this portion of the lab, you will see an accurate representation of living cells.

Procedure: Making a Wet Mount

1. Place a drop of suspension on a slide.
2. Slowly cover the drop with a cover slip by holding the cover slip between thumb and forefinger; lower it until the left edge of the cover slip touches the slide.
3. Then tilt the cover slip at a 40-degree angle to the slide and move it toward the drop until the fluid flows into the acute angle between the slide and the cover slip (see Figure 7-1). Slowly lower the slip onto the drop.

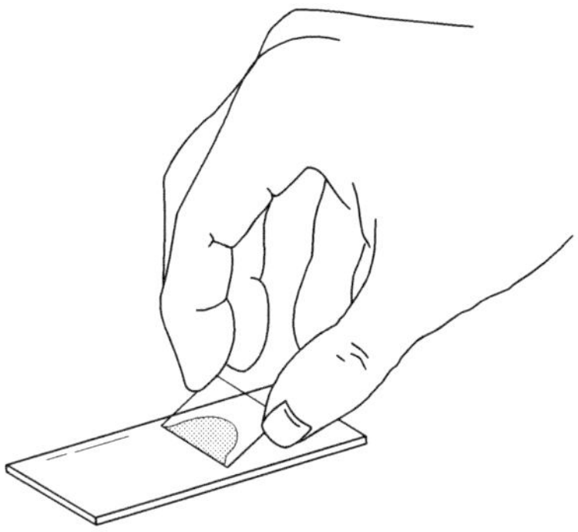

Figure 7-1. Procedure for making a wet mount.

OBSERVING SINGLE-CELLED ORGANISMS

The first group of organisms you will observe belong to the kingdom *Protista*. Members of this group are single-celled eukaryotic organisms. Some of the protists are plantlike, with the ability to manufacture their own food. Other protists are more animal-like; they must ingest their food. You will make observations on a **heterotrophic** (ingests food) protist, *Amoeba,* and an **autotrophic** (manufactures its own food) protist, *Volvox.*

Amoeba

(Phylum *Sarcodina:* amoeboid protists)
Size: 600 μm (1 μm = 1 micrometer = 1 millionth of a meter)

1. Locate amoebas in the culture using the dissecting microscope. They will appear to be a whitish mass. Pipette them onto a slide.
2. Add a drop of Proto-Slo and mix.
3. Slowly cover the drop with a cover slip.
4. View first at low power (10× objective) and then at high power (40× objective), partially closing the iris diaphragm.
5. Label the organism and cell structures (nucleus, cell [plasma] membrane, food **vacuole,** and **contractile vacuole**) that you can see on Figure 7-2.
6. How does an amoeba move?

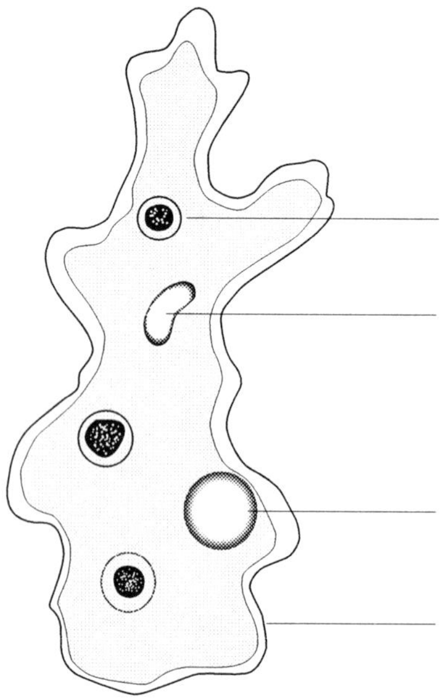

Figure 7-2. Amoeba.

Volvox

(Division *Chlorophyta:* green algae)
Size: 350 μm to 500 μm.

1. Obtain a depression slide. Place a few drops of suspension of *Volvox* into the well of the slide.
2. Cover with cover slip.
3. Observe first under low power (10×) and then high power (40×). Note that *Volvox* is a spherical colony of identical cells.
4. In the space provided, draw a colony and label individual cells and daughter colonies (asexual reproductive structures), chloroplasts, and flagella.

Your drawing of Volvox.

5. Why is the *Volvox* green in color?

6. How does the *Volvox* obtain its food?

OBSERVING FUNGAL CELLS

In this part of the exercise, you will make observations of two common living fungi: yeast (an ascomycete) and a mushroom (a basidiomycete). You will notice that fungal cells have a **cell wall** like the photosynthetic protists and higher plants. In contrast to photosynthetic organisms, fungal cells lack chloroplast; fungi must absorb their food from their environment. Yeast are small organisms that can reproduce asexually by **budding**. In the process of budding, yeast cells divide, but cytokinesis is incomplete. The result is a chain of cells with distinct nuclei. Mushrooms are the fruiting structures of a fungus belonging to the Division *Basidiomycota*. The strands of fungal cells are called **hyphae.** The hyphae of a single fungus are collectively referred to as a **mycelium**. You will make observations of the sexual reproductive spores of the mushroom.

Yeast

(*Saccharomyces*, Division *Ascomycota*)

1. Obtain a clean slide. Place a few drops of suspension of yeast onto the slide.
2. Add one drop of methylene blue.
3. Cover with cover slip.
4. Observe first under low power (10×) and then high power (40×). Note that yeast are spherical to oval in shape. Is there any evidence of budding?
5. In the space provided, draw and label yeast cells including the nucleus, cell wall, and **cytoplasm**.

Your drawing of yeast.

Mushroom

(*Agaricus*, Division *Basidiomycota*)

1. Obtain a clean slide. Take a razor blade and remove a single, thin, dark-brown gill from the underside of the mushroom cap that has fully opened. Place the gill on the slide and add a drop of water.
2. Add one drop of methylene blue.
3. Cover with cover slip.
4. Observe first under low power (10×) and then high power (40×).
5. You should see rows of clear club-shaped saclike structures called basidia. Inside each basidium are the sexual reproductive spores,

basidiospores. In the space provided, draw and label what you see and include the following structures: hypha, basidium, basidiospore, nucleus (of any individual cell you see), cell wall (of any individual cell you observe), and cytoplasm.

Your drawing of mushroom gill.

Questions for Thought

1. What cell structures do the yeast and mushroom cells share with the protists that you observed?

2. How are the yeast and mushroom cells similar to those of the *Volvox?* How are they different?

3. Given that fungal cells lack chloroplast, how do fungi obtain their food?

OBSERVING PLANT CELLS

The cells you will observe in this portion of the lab exercise are plant cells. The cells are derived from different parts of a plant: a fruit (bell pepper), a leaf *(Elodea)*, a tuber (potato), and a root (purple onion). As you make observations of these cells, take note of any special adaptations that are unique to the part of the plant from which the cells were collected. You will look at two types of storage organelles: chromoplasts containing pigment and leukoplasts containing starch. A chloroplast is a specific type of chromoplast; it contains the photosynthetic pigment chlorophyll.

Purple Onion

1. Peel off a very thin portion of the purple onion epidermis and place the section in a drop of water on a clean slide.
2. Carefully place a cover slip on the specimen by putting the cover slip at an angle until it contacts the edge of the water droplet. Drop the cover slip and gently tap any air bubbles out from under the cover slip.
3. Using the 10× and 40× objectives, locate the nucleus, **nucleolus**, nuclear envelope (nuclear membrane), and cell wall.
4. In the space provided, draw and label a few "typical" purple onion cells. Be certain to label the structures.

Your drawing of purple onion cells.

Red Bell Pepper

1. Peel off a portion of the skin and epidermis of the pepper. Try to get a very small portion of the pepper. (*Hint:* If your cover slip is at an angle, the specimen is too thick.)
2. Place the pepper in a drop of water, and place the cover slip on the slide. Remove any bubbles from the slide.
3. View the specimen using the 10× (low) and 40× (high) objectives. Locate the red-colored **plastids**, central vacuole, and cell wall.

4. In the space provided, draw and label the indicated structures.

Your drawing of red bell pepper cells.

Anachoris

1. Obtain a single *Anachoris (Elodea)* leaf from the uppermost part of the stalk (the leaves should be almost transparent).
2. Place the leaf on a flat slide and add a drop of water and a cover slip.
3. Using the eraser end of a pencil, gently tap the cover slip to force out the air bubbles that may have been trapped on the surface of the leaf.
4. Remember that the leaf is several cell layers thick; you need to focus on a cell close to the edge of the leaf where the number of cells is reduced. Focus using the 10× (low-power) and 40× (high-power) objectives.
5. Locate the cell wall, cytoplasm, chloroplasts, and the large central vacuole (chloroplasts are absent from this part of the cell).
6. Observe cytoplasmic streaming in the *Anachoris* leaf—the chloroplasts will appear to be moving in a circle inside the cell close to the cell wall.
7. In the space provided, draw and label what you observe.

Your drawing of the Anachoris (Elodea) leaf cells.

Potato

1. Using a razor blade, cut a very thin section of the potato; be certain that you do not have any of the peel in your section.
2. Place the potato section on a flat slide. Add a drop of water and a drop of Lugol's solution (potassium iodide) and put a cover slip on the specimen. *Note:* If your cover slip is at an angle on the slide, your potato section is too thick; you will need to start over.
3. The Lugol's solution reacts with the starch stored in the leukoplasts. Starch reacts with Lugol's solution and forms a blue-black color.
4. Using the 10× (low-power) and 40× (high-power) objectives, focus on the thinnest portion of your section. The cells should look like slightly irregular boxes containing dark-blue balls (leukoplasts).
5. In the space provided, draw and label the potato cells including the following structures: cell wall, cytoplasm, leukoplasts.

Your drawing of potato cells (starch stained with potassium iodide).

Questions for Thought

1. What other organelle(s) have you observed which are similar in function to the red plastids of the pepper?

2. Which structure(s) are common to the purple onion, red bell pepper, *Anachoris,* and potato?

3. Why are there differences in the basic cell structure of the purple onion/potato, the red bell pepper, and *Anachoris* (*Hint:* What part of the plant does the tissue come from?)

OBSERVING ANIMAL CELLS

Most multicellular animals have bodies containing tissues (a group of cells specialized to perform a specific function). In vertebrates, there are four general types of tissues: epithelial, connective, muscle, and nerve. In this part of the lab exercise, you will make observations of an example of each of these tissues.

Epithelial Tissue—Squamous Cells

The surface tissue lining the inside of the cheek is a sheet of squamous epithelium cells. In humans, epithelial tissues line many of the hollow internal body organs and cover the surface of the body.

1. Gently scrape the inside of your cheek with the flattened end of a toothpick.
2. Make a wet mount by floating the scrapings off into a drop of 0.9% NaCl on a slide. Repeat this procedure four or five times.
3. Place a cover slip on the slide and add a drop or two of methylene blue to one end of the cover slip; draw the stain across the slide by putting a paper towel at the other side of the cover slip. Wipe away any extra stain and view under low and high power.
4. In the space provided, draw your observations, labeling the nucleus, cytoplasm, and plasma membrane (cell membrane).

Your drawing of cheek epithelial cells.

Connective Tissue—Blood

Connective tissues function to connect two different tissue types within an organ, usually an epithelial tissue with some type of muscle. All connective tissues are made of a nonliving substance, a matrix, and living cells within the matrix. The connective tissue you will look at is blood; the nonliving portion of the blood is the plasma (the liquid part). You will look at the three types of cells found in blood erythrocytes (red blood cells), leukocytes (white blood cells), and thrombocytes (platelets). Erythrocytes are small thin cells lacking a distinct nucleus at functional maturity; red blood cells are responsible for transporting gases (O_2 and CO_2) in the blood. Leukocytes are involved in immunity, and the size and shape of their nucleus and the presence or absence of granules in the cytoplasm can distinguish the five different types of white blood cells. Thrombocytes are very small irregular cells involved in blood clot formation.

1. Obtain a prepared slide of human blood that has been stained with Wright's stain. The Wright's stain differentially stains the different types of white blood cells; the platelets are also evident.
2. Begin with the 10× (low-power) and 40× (high-power) objectives. When you are certain that your slide is in focus, place a drop of immersion oil on the cover slip and put the oil immersion (100×) objective in place. Use the fine adjustment knob to make your final focus.
3. In the space provided, draw and label what you see, including erythrocyte, leukocyte, and thrombocyte. Label the nucleus and the cell membrane in each cell.

Your drawing of human blood cells.

Muscle Tissue—Skeletal Muscle

Muscles are responsible for movement in multicellular animals. Higher animals possess three distinct types of muscle tissue: cardiac (associated with the heart), smooth (in hollow internal structures), and skeletal (attached to bones). Skeletal muscle is multinucleate (contains more than one nucleus per cell) and striated (ridged and irregular in surface appearance). The cells are elongated cylinders and are parallel in arrangement.

1. Obtain a prepared slide of skeletal muscle and focus using the 10× (low-power) and 40× (high-power) objectives.
2. In the space provided, draw and label the skeletal muscle cells including the nucleus, cytoplasm, cell membrane, and striations.

Your drawing of skeletal muscle cells.

Nervous Tissue—Neurons

Nervous tissue is composed primarily of specialized cells called neurons. Neurons are relatively large cells with a distinct nucleus and elongated cytoplasmic projections (axon and dendrites) along which nerve impulses travel.

1. Obtain a slide of a human motor neuron that has been stained with a silver preparation.
2. Use the 10× (low power) objective to make your initial focus. You will notice that neurons are much larger cells than the cheek, blood, and muscle cells you have examined. Use the 40× (high-power) objective to look at details in different parts of the neuron.

3. In the space provided, draw and label a neuron including the nucleus, cytoplasm, axon (longest single cytoplasmic projection), dendrites (smaller projections), and cell membrane.

Your drawing of a nerve cell.

Questions for Thought

1. How do the cheek epithelial cells compare with the purple onion cells in shape and size?

2. How do these differences in shape and size relate to the function that each of these cell types perform in the organism?

3. How does the lack of a nucleus in an erythrocyte (red blood cell) relate to its function?

4. If red blood cells lack a nucleus, how do these cells reproduce?

5. Skeletal and cardiac muscle cells have a comparatively large number of mitochondria. Suggest a reason for this phenomenon.

6. How do the shape and size of the neuron relate to its function?

GLOSSARY OF TERMS

autotrophic: (*auto*—self, *trophos*—food) Organisms that use chemical energy (such as sunlight) to make their own food.

budding: A form of asexual reproduction in which outgrowths from the parent pinch off (as in the invertebrate *Hydra*) or remain attached to form extensive colonies (such as in yeast).

cell (plasma) membrane: The outermost membrane of the cell; common to all prokaryotic and eukaryotic cells.

cell wall: A rigid layer in plants, bacteria, some protists, and some fungi surrounding the cell's cytoplasm and cell (plasma) membrane.

chloroplast: (*chloros*—green, *plastos*—molded)—An organelle surrounded by a double-membrane containing extensive internal membranes and chlorophyll; site for photosynthesis in autotrophic eukaryotes.

cilium: (pl. *cilia*) A short, hairlike projection from the cell surface composed of protein microtubules used for locomotion; usually present in large numbers or in a pattern on the surface of cells bearing cilia.

contractile vacuole: a structure present in freshwater protists that collects water coming into the cell and pumps the water back into the environment. Because freshwater protists live a hypotonic environment, water is always moving into the protist's cell; the contractile vacuole maintains proper water balance and prevents the cell from bursting.

cytoplasm: the portion of a eukaryotic cell outside the nucleus of the cell and inside the cell (plasma) membrane.

eukaryotic: (*eu*—true, *karyon*–kernel) Cells which have a membrane-enclosed nucleus and membrane-bound organelles such as chloroplasts and mitochondria. The chromosomes of eukaryotic cells have proteins associated with the DNA, and eukaryotic ribosomes are larger in size than those found in bacteria and cyanobacteria. Animals, plants, fungi, and protists all have eukaryotic cells.

flagellum: (pl. *flagetta*) A long, whiplike extension from the cell's surface composed of protein microtubules found in eukaryotic and bacterial cells used for locomotion; usually present in limited numbers in contrast to cilia.

heterotrophic: (*hetero*—different; *trophos*—food) Organisms that cannot make their own food and must ingest energy-containing molecules produced by other organisms.

hypha: (pl. *hyphae*) (*hyphos*—web, to weave) A multicellular fungal filament that makes up the body of a fungus.

mitochondrion: (pl. *mitochondria*) (*mitos*—thread; *chondrion*—smail grain) A bacterium-like eukaryotic organelle containing a complex internal membrane system where oxidative respiration occurs.

mycelium: (pl. *mycelia*) The dense and branched network of hyphae in a fungus.

nucleolus: A eukaryotic organelle that is the site of RNA synthesis.

nucleus: A membrane-bound eukaryotic organelle that contains the genetic material of the cell (chromosomal DNA).

organelle: A specialized structure within a cell that performs a specific function; organelles are present in both prokaryotic and eukaryotic cells.

plastid: A eukaryotic plant organelle which is surrounded by a double-membrane that is involved in pigment storage or food storage; a chloroplast is a plastid that stores the pigment chlorophyll and is the site of photosynthesis.

prokaryotic: (*pro*—first, *karyon*—kernel) Cells that lack a nucleus enclosed within a membrane and membrane-bound organelles.

vacuole: A single-membrane enclosed vesicle that may contain fluid, wastes, or food in eukaryotic cells.

EXERCISE 8

Genes: Patterns of Inheritance

OBJECTIVES

Upon completion of this lab exercise, you should be able to:

- distinguish between an allele and a gene.
- apply the terms *dominant* and *recessive* to Mendel's Law of Dominance.
- differentiate between the genotype and the phenotype of an individual.
- write the heterzygous and the homozygous genotypes given an allele pair.
- determine the different **gametes** produced by an organism.
- distinguish between an autosomal and a sex-linked gene.
- complete a Punnet square for a monohybrid cross and predict the F_1 genotypic and phenotypic ratios.

Suggested Reading:
Chapter 4 of *The Science of Agriculture: A Biological Approach*, 4th edition

INTRODUCTION

The science of **genetics** deals with patterns of inheritance from one generation of organisms to the next. Early studies of garden peas performed by Gregor Mendel (1822–1884) provided the foundation for understanding the inheritance of traits controlled by a single **gene**. Mendel suggested that genes were discrete units of inheritance; modern research demonstrates that genes are sequences of bases in chromosomal DNA (deoxyribonucleic acid) that code for a single trait. Different forms of the same gene, **alleles**, account for the variation observed by Mendel in pea seed color, seed shape, flower color, and plant height. Gregor Mendel discovered that if he crossed two plants that bred true for alternate forms of a gene, yellow seeds and green seeds, for example, the trait appearing in the first (F_1) generation is the dominant allele. In Mendel's experiment, all of the pea seeds in the F_1 generation were yellow. Yellow is the dominant allele of the seed color gene. Mendel crossed two F_1 yellow seeded pea plants, and a small proportion of the offspring in the next generation had green seeds. The trait reappearing in the second (F_2) generation is the recessive trait: the green seed color in peas. This principle is called Mendel's **Law of Dominance**. By convention, geneticists always use the first letter of the dominant trait to represent the gene. In Mendel's experiment, the letter *Y* would represent the gene for seed color. The dominant allele is always written with a capital letter, and the recessive allele is the lowercase counterpart. The allele for a yellow seed is written as *Y,* and the allele for a green seed is designated by *y*.

The **genotype** is the specific arrangement of alleles. If both alleles for a trait are the same, *YY* or *yy,* the genotype of the individual is homozygous for the trait. If the alleles for a trait are different, *Yy* for example, the individual is heterozygous for the trait (*Note*: In heterozygous genotypes, the dominant allele, represented by a capital letter, is always written first.) The **phenotype** is the outward expression of the genotype. If one dominant allele is present, the dominant allele will be expressed. In pea plants that are homozygous dominant *(YY)* or heterozygous *(Yy)* for seed color, all the seeds are yellow; yellow is the phenotype. For the recessive phenotype to be expressed, the individual must be homozygous recessive *(yy,* green seed color).

One of the classic tools of genetic research is *Drosophila melanogaster,* the fruit fly. The patterns of inheritance for many physical (morphological) and biochemical traits have been identified and associated with specific regions on one of the four *Drosophila* chromosomes. *Drosophila* has a short generation time; at 25°C, fruit flies will produce a new generation in as few as 10 days. The fruit fly will spend 1 day as an egg, 5 days as a larva (**maggot**), and 4 days as a pupa before emerging as an adult (see Figure 8-1).

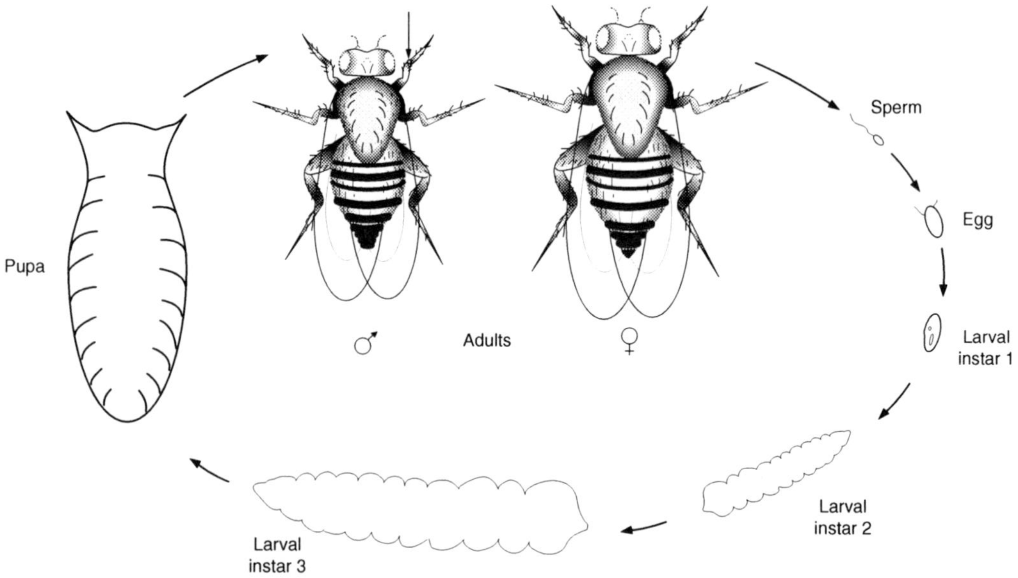

Figure 8-1. The life cycle of the Drosophila melanogaster, the fruit fly.

The combination of rapid reproduction and chromosome maps marking the location of specific genes make *Drosophila* an ideal tool for further genetic research. In addition to Mendel's peas and the fruit fly, several other organisms with desirable characteristics of short generation time and easy culture have also been well studied. The fungus *Neurospora,* the green alga *Cklamydomonas,* the nematode *Caenorhabditis elegans,* the zebra fish *Dania rerio,* the common wall cress *Ara-bidopsis fhalinia,* and the mouse are all excellent models for studying inheritance patterns.

All fruit flies have identical chromosomes 2, 3, and 4. These are **autosomal chromosomes**. Chromosome 1 is an X or a Y. XX is a female and XY is a male *Drosophila.* Because chromosome 1 determines the sex of the fruit fly, it is a **sex chromosome**. Alleles for genes located on autosomal chromosomes are called autosomal alleles. In contrast, genes located on the X chromosome are sex-linked traits. The Y chromosome contains few functional genes (genes expressed as phenotypes), so geneticists generally ignore it. Recessive sex-linked alleles are expressed more frequently in male flies because they have a single X chromosome than in females who have two X chromosomes. For a sex-linked recessive trait to be expressed in a female, two copies of the recessive allele, one on each X chromosome, must be present. Examples of human sex-linked traits are red-green color blindness and hemophilia.

In agriculture, many desirable traits in crops and livestock are managed through selective breeding. Hornless (polled) cattle, nematode-resistant tomatoes, and cucurbit wilt–resistant squash and cucumbers are examples of desirable genetic traits. Most commercially important crop plants were developed through the long process of selective breeding. Beef and dairy cattle producers maintain detailed breeding records for their herds to help maximize the expression of desirable traits in their calves. In this exercise, you will investigate the inheritance patterns of two different corn genes, seed color and sugar content.

DETERMINING THE PATTERN OF INHERITANCE OF CORN SEED COLOR

In this part of the lab exercise, you will determine which corn seed color—purple or yellow—is dominant. You will ascertain which seed sugar content allele is dominant. If two genetic traits always occur together, then the traits are located close to one another on the same chromosome. These traits are **linked**. The inheritance of one linked trait is not independent of the inheritance of another linked trait. You have several ears of corn with purple and yellow seeds on the lab table in front of you. There are also starchy (plump and round) seeds and sweet (shriveled) seeds on the corn ears. You have several questions to answer in this portion of the lab exercise. Using F_1 corn ears, which allele for seed color is dominant and which allele for sugar content is dominant? What were the phenotype and genotype of the parent plants?

Seed Color Questions

1. Given two alleles for corn seed color—*purple* and *yellow*—if you bred homozygous corn plants (one purple and one yellow), what would be the relative distribution of genotypes among F_1 offspring? Show your work.

2. Assume that purple seed color is dominant. What phenotypic ratio would you expect in the F_1 corn plants from a cross of homozygous purple and homozygous yellow corn plants?

3. Assume that yellow seed color is dominant. What phenotypic ratio would you expect in the F_1 corn plants from a cross of homozygous purple and homozygous yellow corn plants?

Seed Color Hypothesis

Write a hypothesis describing the inheritance pattern of corn seed color (purple and yellow alleles).

Seed Color Prediction

Predict the offspring resulting from your hypothesized parental genotypes. (*Hint:* Draw a Punnet square.)

Seed Sugar Content Questions

1. Given two alleles for corn seed sugar content—*starchy* and *sweet*—if you bred homozygous corn plants (one starchy and one sweet), what would be the relative distribution of genotypes among F_1 offspring? Show your work.

2. Assume that sweet corn is the dominant allele for corn seed sugar content. What phenotypic ratio would you expect in the F_1 corn plants from a cross of homozygous starchy and homozygous sweet corn plants?

3. Assume that starchy corn is the dominant allele for corn seed sugar content. What phenotypic ratio would you expect in the F_1 corn plants from a cross of homozygous starchy and homozygous sweet corn plants?

Corn Seed Sugar Content Hypothesis

Write a hypothesis describing the inheritance pattern of corn seed sugar content (starchy and sweet alleles).

Corn Seed Sugar Content Prediction

Predict the offspring resulting from your hypothesized parental genotypes. (*Hint:* Draw a Punnet square.)

Procedures

1. You should work with a lab partner for this portion of the lab exercise. Each partner should obtain an F_1 corn ear from your instructor. Pick a random location on the ear and start counting the number of purple and yellow corn seeds.
2. Use Table 8-1 to identify and record the seed color for 50 corn seeds. Using a different corn ear, your lab partner should identify and record the seed color for 50 corn seeds in Table 8-1 in his or her lab manual. When you analyze your results, you will pool your data, giving you a total sample of 100 seeds (this makes data interpretation easier).

3. Using the same F_1 corn ears, pick a random location on the ear and start counting the number of starchy and sweet corn seeds. The starchy corn seeds are rounded and plump with a smooth seed surface. In contrast, the sweet corn seeds have a shriveled surface. Figure 8-2 illustrates three of the four different corn seed phenotypes.
4. Record your data for 50 starchy and sweet kernels in Table 8-1. Your lab partner should record the numbers of starchy and sweet corn seeds for a second ear of corn in Table 8-1 in his or her lab manual. You will combine your data so that your lab group has a sample of 100 seeds to analyze.

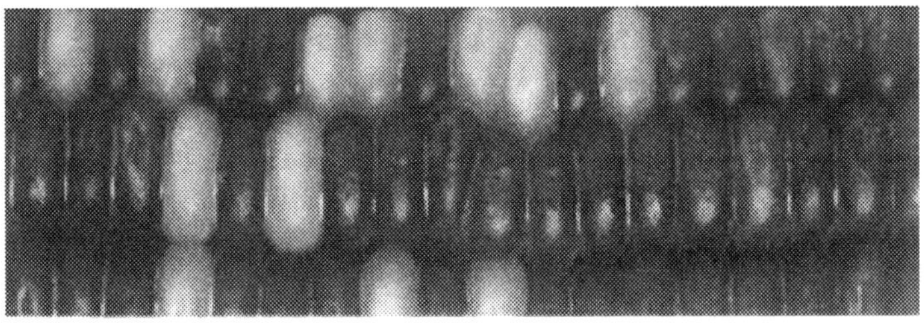

(a) Monohybrid cross: *purple* and *yellow alleles*

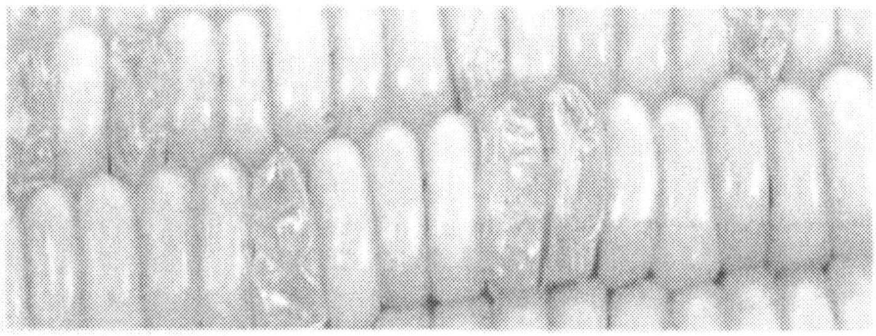

(b) Monohybrid cross: *starchy* and *sweet* alleles

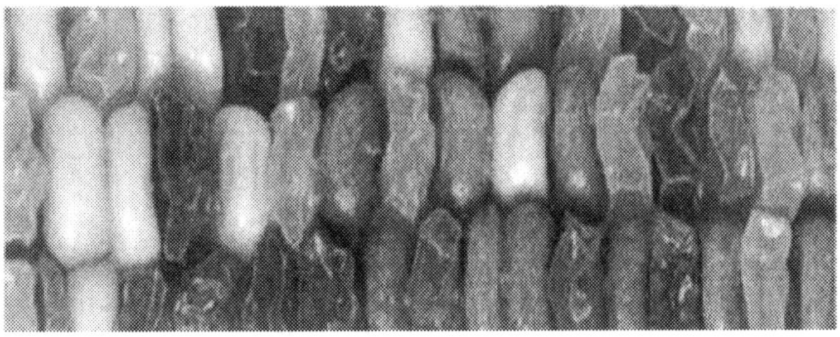

(c) Dihybrid cross: *purple, yellow, starchy,* and *sweet* alleles

Figure 8-2. The three of the four different corn seed phenotypes.

Seed. no	Seed color (purple/ yellow)	Sugar content (starchy/ sweet)	Seed. no	Seed color (purple/ yellow)	Sugar content (starchy/ sweet)
1			26		
2			27		
3			28		
4			29		
5			30		
6			31		
7			32		
8			33		
9			34		
10			35		
11			36		
12			37		
13			38		
14			39		
15			40		
16			41		
17			42		
18			43		
19			44		
20			45		
21			46		
22			47		
23			48		
23			49		
25			50		

Table 8-1. F_1 offspring of cross of homozygous purple and yellow and homozygous starchy and sweet corn plants

Data Interpretation and Results

1. Going back to Mendel's Law of Dominance and using the data in Table 8-1, which allele is dominant for corn seed color?

2. Going back to Mendel's Law of Dominance and using the data in Table 8-1, which allele is the dominant allele for seed sugar content, starchy or sweet?

3. Explain the evidence you have to support your assertion. How many total seeds (out of 100) are purple as compared with yellow?

   How many are starchy as compared with sweet?

4. Let's name the gene. For any gene with two alleles, there is a wild-type form, which is the most common form in nature, and a mutant form. The mutation is a change in the chromosomal DNA in the region of the gene. If the mutation is dominant, capitalize the name. If the mutation is recessive, use lowercase letters for the name. Based on your observations, should the yellow allele be written with capital or small letters?

   What about the sweet allele?

5. Do the data support or fail to support your hypothesis regarding the inheritance of corn seed color (purple or yellow)?

6. Do the data support or fail to support your hypothesis regarding the inheritance of corn seed sugar content (starchy or sweet)?

7. What conclusions can you draw about these alleles and their inheritance patterns?

PARENTAL GENOTYPES AS A FUNCTION OF F_2 OFFSPRING

Now that you have determined the inheritance patterns for corn seed color and sweetness, it is possible to determine the genotypes and phenotypes of the parents by making observations of the offspring. You will be given F_2 corn ears resulting from a dihybrid cross of parents that have the purple or yellow corn seed allele and the starchy or sweet corn seed allele. Not all class members will have the ears from crosses of the same parents. Your instructor will assign you to lab groups who will work on corn ears resulting from the same cross.

Procedures

1. You should work with a lab partner for this portion of the lab exercise. Each partner should obtain an F_2 corn ear from your instructor. Pick a random location on the ear and start counting the number of purple and starchy, purple and sweet, yellow and starchy, and yellow and sweet corn seeds.
2. Use Table 8-2 to identify and record the seed color and sugar content for 50 corn seeds. In Table 8-2, record the seed color in the first column (purple or yellow) and the seed sugar content (starchy or sweet) in the second column beside the seed number. Using a different corn ear, your lab partner should identify and record the seed color and sugar content for 50 corn seeds in Table 8-2 in his or her lab manual. When you analyze your results, you will pool your data, giving you a total sample of 100 seeds (this makes data interpretation easier).

Seed. no	Seed color (purple/ yellow)	Sugar content (starchy/ sweet)	Seed. no	Seed color (purple/ yellow)	Sugar content (starchy/ sweet)
1			26		
2			27		
3			28		
4			29		
5			30		
6			31		
7			32		
8			33		
9			34		
10			35		
11			36		
12			37		
13			38		
14			39		
15			40		
16			41		
17			42		
18			43		
19			44		
20			45		
21			46		
22			47		
23			48		
23			49		
25			50		

Table 8-2. F_2 offspring from a cross of unknown corn plants (purple or yellow and starchy or sweet alleles)

Data Interpretation and Results

Table 8-3 summarizes the total number of corn seeds for each corn seed phenotype (from Table 8-2):

Phenotype	Number observed	Experimental group total
Purple and starchy		
Purple and sweet		
Yellow and starchy		
Yellow and sweet		
Totals		

Table 8-3. Summary of F_2 corn seed phenotypes for individual and lab experimental group data

Let's summarize the earlier experiments:

1. Which allele, *purple* or *yellow,* is dominant?

2. Which allele, *starchy* or *sweet,* is dominant?

Hypothesis

Using the observations that you and your lab group have made regarding the inheritance of corn seed color and corn kernel sugar content, write a hypothesis about the parental genotypes that produced the F_2 corn kernels.

Prediction

Predict the outcome of the experiment (use an "If/then" statement).

Hypothesis Testing

You will use the class data summarized in Table 8-3 to determine whether the data support or fail to support your hypothesis using a chi-square goodness of fit test.

A Word about the Chi-Square Goodness of Fit

One of the objectives of statistics is to make inferences about a population (in this case, all of the corn seeds from all corn ears of the same genetic cross) from a sample (the 100 corn seeds that you and your lab partner counted). In the case of the Yellow *(y)* and Sweet *(su)* alleles, you have determined the patterns of inheritance. Given what we know about Mendel's laws and the parental genotypes of the corn involved, we can make predictions about the relative proportion of individuals who inherit the yellow (wild-type) and purple (mutant) alleles and the sweet (wild-type) and starchy (mutant) alleles. The chi-square statistic compares the number of expected wild-type and mutant phenotypes with the actual number observed (from the class data). Chi-square results indicate whether the difference between the expected and observed phenotypes is outside the realm of random chance. The chi-square statistic is calculated using the formula:

$$X^2 = \Sigma \frac{(O - E)^2}{E}$$

Note: O = Observed frequency, E = Expected frequency

Let's look at an example of how the chi-square test can be used. Suppose we want to examine the relative number of male and female babies born in Atlanta, Georgia, the second week in December 1993 (9 months after a big winter storm). We expect that the probability of conceiving a male child is ½ or 50%, and we expect that the probability associated with conceiving a female child is also ½ or 50%. If 840 children were born during the second week of December, how can we use the chi-square statistic to determine whether the relative proportion of males to females differs significantly from 1 to 1? Using the data in Table 8-4, we can calculate X^2 for this problem.

We know that 840 children were born. If the probability of having a female child is ½ (50%), we obtain the expected number by multiplying ½ (the probability) by 840 (the total number).

Expected number of female children = ½ × 840 = 420 female

The same calculation is done for the male children. *Hint:* The total number of expected children should equal the total number of observed births if you have done your calculations correctly.

Gender	Number of births Observed	Expected
Female	324	420
Male	516	420
Totals	840	840

Table 8-4. Data for chi-square test

$$X^2 = \Sigma\frac{(O-E)^2}{E} = \frac{(324-420)^2}{420} = \frac{(516-420)^2}{420}$$
$$= 21.9429 + 21.9429 = 43.8857$$

Next we need to determine the degrees of freedom (df) for the test. The degrees of freedom is equal to the number of different categories observed minus one. In this example, we have two categories: female and male. Thus, the degrees of freedom are 2–1 or 1. To use an X^2 table, we must also determine the α (acceptable error rate, α). For the purposes of our experiments, assume that α equals 0.05 (5%). Using the chi-square in Table 8-5, the critical X^2 with 1 df and $\alpha = 0.05$ is 3.84146. If the calculated value of X^2 is less than 3.84146 (critical value), then we fail to reject our hypothesis and conclude that there were no differences in the number of male children as compared with female children born during the second week of December 1993 in Atlanta, Georgia. If the calculated value of X^2 is greater than or equal to the

Allele	Observed value (O)	Expected value (E)	(O-E) Deviation	$(O-E)^2$ Deviation^2	$\frac{(O-E)^2}{E}$
Yellow or sweet (wild type)					
Purple or starchy (mutation)					
$X^2 = \Sigma\frac{(O-E)^2}{E}$ critical X^2 ($\alpha = 0.05$, df = ___) = ___					$\Sigma =$

Table 8-5. Chi-square test of hypothesis for corn seed color or sugar content alleles

critical value, 3.84146, then we reject our hypothesis and conclude with 95% certainty (100%—our error rate) that there was a statistically significant difference between the expected numbers of male versus female children and the observed numbers of male and female babies born during the second week of December 1993. Because the calculated value of X^2, 43.8857, is greater than the critical value, 3.84146, we reject the hypothesis and conclude that the relative proportion of male and female children born during the second week of December 1993 in Atlanta, Georgia, is statistically different from what would be expected as a result of random chance.

To complete a test of hypothesis on the yellow *(y)* and purple *(Y)* alleles or sweet *(su)* and starchy *(SU)* alleles, you will need to determine the probabilities associated with inheriting each of the alleles, the correct degrees of freedom (df), the critical value of chi-square ($\alpha = 0.05$), and calculate the X^2 statistic from the class data using the formula. Compare the calculated X^2 statistic with the critical value of X^2 for 5% error and degrees of freedom (chi-square table follows this exercise) to complete the test of your hypothesis.

Data Interpretation and Results

1. Calculate the expected values for each trait based on the total number of seeds counted in your class (use the data in Table 8-3).

2. Do the class results require you to reject or fail to reject your hypothesis? Why?

3. Does this experiment support or contradict your conclusions regarding the inheritance of the yellow or sweet alleles?

4. What are the implications of your class findings for large-scale producers of corn if they are planting varieties that self-pollinate (i.e., self-cross)?

5. Would the qualities of a harvested corn crop be affected by the degree of heterozygosity of the corn seed?

6. How would the pollination mechanism (self-pollination vs. cross-pollination) affect the inheritance of desirable traits in a large cornfield?

GLOSSARY OF TERMS

allele: (*attelon*—of one another) Alternate forms of the same gene. For example, *Y* codes for yellow pea seed color, whereas *y* codes for green pea seed color. *Y* and *y* both determine seed color, but each results in a different expression of color.

autosomal chromosomes: (*autos*—self, *soma*—body) A eukaryotic chromosome that is not involved in sex determination.

gametes: (*gamein*—to marry) Reproductive cells that are haploid (N) in animals and plants.

gene: (*genos*—offspring) A sequence of bases in chromosomal DNA (deoxyribonucleic acid) that codes for a single trait.

genetics: The study of patterns of inheritance among organisms.

genotype: (*genos*—offspring, *typos*—form) The actual arrangement of alleles determining the genetic composition of an individual.

Law of Dominance: If two individuals who are homozygous for alternate forms of the same gene are crossed, the trait that is expressed in the F_1 generation is the dominant allele. Subsequently, if two F_1 individuals are crossed, the trait which reappears in the next generation is recessive.

linked: Located on the same chromosome; linked genes are always segregate together.

phenotype: (*phaeinein*—to show, *typos*—form) The outward expression of the genotype (specific alleles for a trait).

principle of segregation: Gregor Mendel proposed that when gametes are formed via meiosis, two alleles for a given gene separate so that each gamete contains a single allele for the trait.

sex chromosomes: Chromosomes involved in the determination of sexual characteristics. In mammals, the X and Y chromosomes are the sex chromosomes.

EXERCISE 9

Recombinant DNA Technology: Plasmid Transformation

OBJECTIVES

Upon completion of this lab exercise, you should be able to:

- describe the structure and uses of a plasmid.
- distinguish plasmid DNA from chromosomal DNA.
- name the characteristics of plasmid that make them useful in bacterial transformation.
- define transformation.
- list factors that affect transformation efficiency.
- count the number of bacterial colonies and convert the bacterial count into the number of cells/milliliter.
- calculate transformation efficiency.
- use sterile technique in the transformation experiment.
- describe the changes in the *Escherichia coli* cells necessary for transformation to occur.
- explain the purpose of the use of Luria broth and Luria broth + kanamycin plates in the experiment.
- explain the advantage of transforming a kanamycin-resistant gene in addition to a gene of interest.

Note:

This lab exercise contains advanced laboratory techniques and presumes some knowledge of DNA replication, RNA transcription, and protein translation. This lab is suitable for inclusion in an advanced placement or gifted program curriculum.

Suggested Reading:

Chapter 5 of *The Science of Agriculture: A Biological Approach*, 4th edition.

Safety Note

Students who are immuno-compromised should use caution when handling the bacterial cultures.

INTRODUCTION

Scientists working in research laboratories have developed the ability to alter the genetic makeup of some organisms. Some of the techniques used by DNA scientists were applications of experimental results from work with bacteria and the **bacteriophages** (viruses) that infect bacterial cells. In the late 1960s, several teams of scientists were able to isolate enzymes from the bacterium, *Escherichia coli* (*E. coli*), that destroyed viral DNA. Further research suggested that the bacteria were able to protect their own DNA by chemically altering the structure of the DNA (methylation). The bacterial enzymes work only on unaltered phage (viral) DNA. These bacterial enzymes, **endonucleases**, search for specific base pair sequences, usually four or six nucleotides (base pairs) long, cut the DNA at a specific point, and are named for the organism from which they are derived.

In 1972, the first recombinant DNA experiments demonstrated that DNA molecules could be altered. Dr. Paul Berg used the restriction endonuclease *EcoRI* to cut a monkey virus and a plasmid. Berg joined the two pieces of DNA with an enzyme, **DNA ligase**, which acted as a chemical "glue," catalyzing the reaction that recombined the monkey virus and plasmid DNA into a single hybrid circular molecule (Micklos and Freyer, 1987). **Plasmids** are small circular pieces of DNA that are separate from the main bacterial chromosome, and capable of independent replication. Plasmids are relatively small molecules ranging in size from 1,000 to 200,000 base pairs in length. Plasmids are useful tools to the molecular scientist because plasmids can contain accessory genes that are expressed by the bacterium. Plasmids are also **vectors**; they carry foreign (nonhost) DNA sequences. Plasmids can transfer these DNA sequences from one bacterial host to another. The foreign DNA is replicated in the host bacterial cell separately from the bacterial chromosome. One example of genes that are frequently carried by bacterial plasmids are genes for antibiotic resistance. Since human and animal diseases that are bacterial in origin are treated with antibiotics, it is an enormous advantage for a bacterium to develop antibiotic resistance; the bacterium survives the course of antibiotic treatment. Furthermore, antibiotic resistance carried on plasmids spreads rapidly through a bacterial population. This trend is found in human populations when antibiotics are overprescribed; the appearance of antibiotic-resistant strains of streptococcus, tuberculosis, and syphilis are due to genes carried on bacterial plasmids.

Molecular scientists can move genes of interest between organisms using plasmids as vectors. **Transformation** occurs when DNA is transferred, assimilated, and expressed by a cell. Transformation of cells happens in nature at a very inefficient rate. In 1970, Mandel and Higa discovered that the ability of *E. coli* cells to take up a plasmid is greatly enhanced by treatment in a cold calcium chloride solution followed by a heat shock at 42°C. Cells with increased capacity for transformation are called **competent cells**. Bacterial cells that are actively dividing (at log phase of growth cycle) and subjected to the Mandel and Higa protocol are more competent than cells in other phases of the growth cycle. The cold calcium chloride solution affects the structure of the bacterial cell wall rendering it porous. The heat shock treatment alters the chemical structure of the bacterial enzymes so that they are nonfunctional. Thus, the foreign DNA introduced during the transformation process is not destroyed by the bacterial cell's battery of protective enzymes.

In this lab exercise, you will be given some *E. coli* cells that are competent for transformation. You will treat the competent bacterial cells with a solution containing a known amount of *p*KAN (a plasmid containing a gene for kanamycin resistance). You will then determine how many of the cells were transformed by growing the cells on several different nutrient media—one of which contains kanamycin. Only the *E. coli* cells that contain the *p*KAN plasmid will survive on media with kanamycin.

Preparation

- Each lab team will require a 500 pi aliquot of competent cells in a sterile 15 ml culture tube.
- Adjust water bath to 42°C.
- Prewarm incubator to 37°C.

Supplies and Equipment

500 pi competent *E. coli* cells (ice cold)
0.015 pg/pl *p*KAN (ice cold)
6 agar plates, as follows:
3 Luria broth
3 Luria broth + kanamycin
Luria broth (sterile)
- crushed ice
- 95% ethanol
- permanent marker
- 250 ml beaker

microfuge
sterile micropipet tips
sterile 15 ml culture tubes
spreading rod
0.5–10 µl micropipet
100–1,000 µl micropipet
- bunsen burner
- ice bucket
- timer
- 37°C and 42°C water baths

NOTE:

In order to obtain accurate results, this entire laboratory exercise must be done using a **sterile technique**. Wipe your lab bench down with the disinfectant provided by your instructor. Be certain to use sterile culture tubes and micropipet tips. It is also important that you follow the time guidelines exactly in order to obtain acceptable transformation efficiency.

TRANSFORMATION WITH *p*KAN

The first phase of this procedure involves the transformation of competent cells with the kanamycin-resistant plasmid (*p*KAN). This portion of the lab will require approximately 45 minutes for completion. Follow the outlined procedure.

1. Mark two sterile culture tubes as follows:

   Tube X—contains *p*KAN

   Tube Y—*p*KAN absent
2. Use 100–1,000 pi micropipet *with sterile tips* to transfer 200 pi of suspended cells each into tubes X and Y.
3. Place tubes on ice.
4. Use 0.5–10 pi micropipet to add 10 pi of *p*KAN solution to tube X.
5. Close cap of tube X and mix DNA with cells by gently swirling suspension or tapping lightly on the bottom of the culture tube.
6. Return both tubes to an ice bucket for 20 minutes.
7. Remove the culture tubes from ice, and "heat shock" cells by *immediately* placing tubes in 42°C water bath for 90 seconds. To ensure "heat shock," the cells must experience a sharp temperature change.
8. Remove tubes from water bath, and *immediately* return to ice beaker for 1 minute.
9. Use 100–1,000 pi micropipet to add 800 pi of Luria broth to each tube. Close the culture tube cap and gently swirl the suspension to mix.
10. Incubate the cells, with moderate agitation in 37°C shaking water bath for 20 to 60 minutes. If a shaking water bath is unavailable, incubate the tubes in a plain water bath. Provide handshaking at regular intervals, while the tubes remain immersed. This will improve transformation results.

SPREADING PLATES

The next step in this experiment involves spreading the transformed *E. coli* colonies on Luria broth plates. Half of the culture plates will contain Luria broth to which kanamycin has been added. On these plates, only *E. coli* colonies containing the *p*KAN plasmid will grow. To obtain well-defined single bacterial colonies on the nutrient plates, a serial dilution technique is used. This involves the successive dilution of a solution of known concentration. This process should require approximately 10 to 15 minutes for completion.

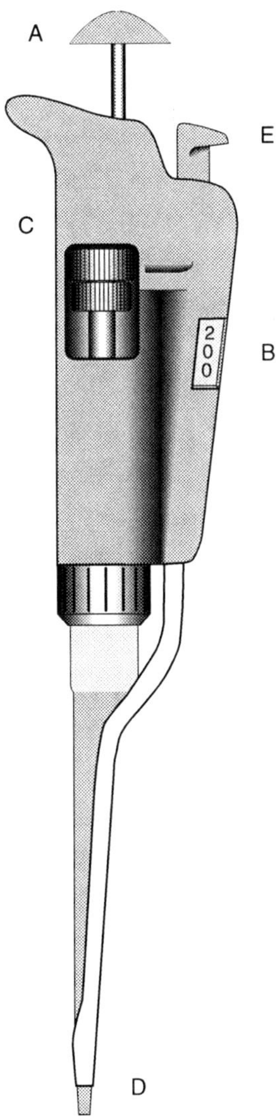

Figure 9-1. Directions for use of a pipetman.

1. The size of the Pipetman (in microliters) is indicated on the top if the plunger (A): 1,000 µl (blue), 200 µl (yellow), and 20 µl (yellow).
2. The volume to be dispensed is displayed in window (B), and set by turning knob (C). Your instructor will preset the amount for this lab exercise.
3. Obtain a sterile tip by opening the box containing tips and inserting the Pipetman (D) into a tip (color-coded to Pipetman).

   To avoid contamination, DO NOT TOUCH THE TIPS, THEY ARE STERILE.
4. Plunger (A) has two stops. The first is for uptake of liquid and the second is for dispensing.
5. TO TAKE UP LIQUID: Push plunger (A) to first stop. Insert the Pipetman into the liquid. Slowly release the plunger, avoiding air bubbles.
6. TO DISPENSE LIQUID: Push plunger (A) to second stop, all the way down past resistance of the first stop.
7. To remove the disposable tip, push small plunger (E).

1. Set up a matrix as illustrated in Table 9-1 to use as a checklist in the plating process (after CSH, *DNA for Beginners,* 1987).
2. Mark the three Luria broth + kanamycin plates: A, B, and C. Mark the three plain Luria broth plates: D, E, and F.
3. Use the micropipet to add 100 pi of Luria broth to plates A and D. The Luria broth will aid in spreading the small volume of cell suspension across the agar plates.
4. Use the 0.5–10 pi micropipet to add 10 pi of cell suspension from Tube X to plates A and D.
5. Use the 100–1,000 pi micropipet to add 100 pi of cells from Tube X to plates B and E.
6. Use the 100–1,000 pi micropipet to add 100 pi of cells from Tube Y to plates C and F.
7. To sterilize the spreading rod, follow this procedure (see Figure 9-2):
   A. Dip the spreading rod in ethanol, and pass it through a Bunsen flame to ignite the alcohol.
   B. Remove the spreading rod from the flame and allow the alcohol to burn off away from the Bunsen flame. (*The spreading rod will become too hot if you leave it in the Bunsen flame.)*
   C. Cool the rod by rubbing it on the side of the agar plates away from the cells, or by rubbing it on the condensed water on the plate lid. *(It is essential to cool the rod before spreading the bacterial cells—if the rod is too hot, the cells will be killed.)*
8. Lift the top of the first plate only enough to perform the following procedure.

   To ensure sterile procedure, do not set the top down on the lab bench.

   A. Spread the cells evenly over the plate surface by moving the bent glass rod back and forth several times over the agar surface.

Plate	Luria broth	X (*p*KAN)	Cells (no *p*KAN)
A	100 µl	10 µl	___
B *p*KAN	___	100 µl	___
C	___	___	100 µl
D	100 µl	10 µl	___
E LB	___	100 µl	___
F	___	___	100 µl

Table 9-1. Setup for transformation of *E. coli* with *p*KAN.

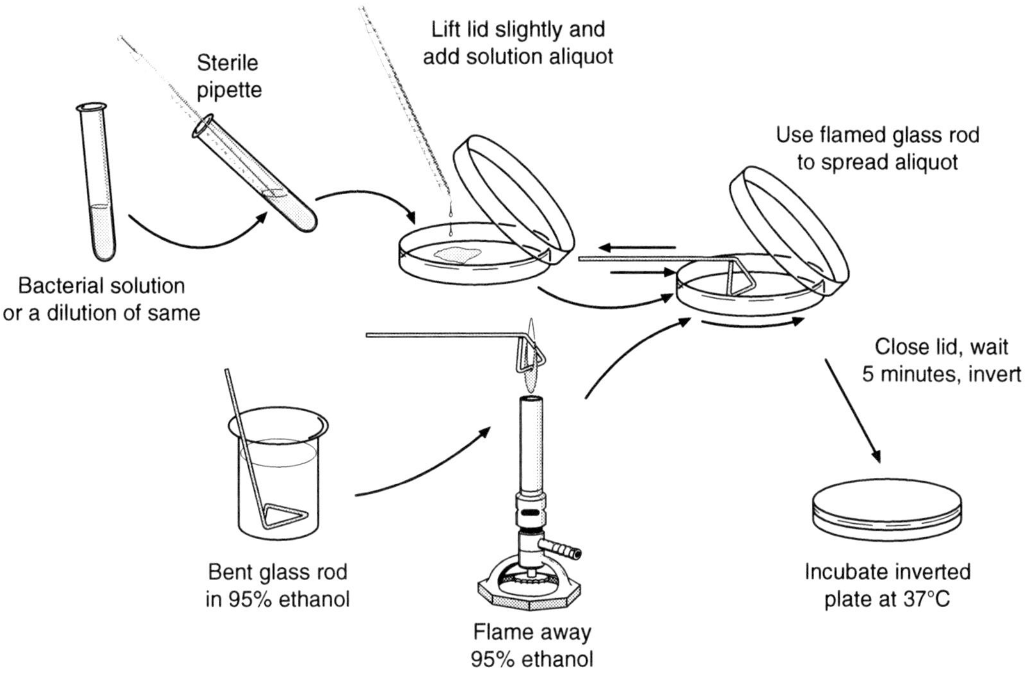

Figure 9-2. Procedure for sterilizing spreading rod and piating ceiis.

B. Rotate the plate *V2* turn and repeat the spreading motion.

C. Replace top.

9. Sterilize the spreading rod and repeat the spreading technique successively for each of the remaining plates.
10. Invert the plates and incubate at 37°C for 12 to 24 hours.
11. The remaining observations should be made 12 to 24 hours after the plates have been spread. After 24 hours, the plates should be refrigerated to retard contaminant growth.

Results

1. Make a rough count of colonies per plate and record the information in Table 9-2. Do not count tiny "satellite" colonies, which emerge from the edges of large, well-established colonies.

   If there are too many colonies to count individually, use a permanent marker to divide the plate into quadrants (either four or eight quadrants depending on colony density). Count the colonies in one quadrant and multiply the number of quadrants to calculate the total number of colonies.
2. Were your results as expected? Explain possible reasons for variations from the expected results.

Transformed cells Plate	(X) 10 µl *p*KAN	(Y) 100 µl *p*KAN	No *p*KAN
Luria broth + kanamycin	A =	B =	C =
Luria broth only	D =	E =	F =

Table 9-2. Results of transformation experiment.

3. Transformation efficiency is expressed as the number of antibiotic-resistant colonies per µg *p*KAN DNA. To determine the transformation efficiency:

   A. Using the concentration of stock *p*KAN as a starting point, determine the mass of *p*KAN (in µg) used in steps 1–4, "Spreading Plates."
   B. Next, calculate the concentration of *p*KAN (in µg/µl) in the total suspension of cells + Luria broth from steps 1–2 and 1–9, "Spreading Plates."
   C. Calculate the mass of *p*KAN (in µg) in the 100 1 of cell suspension spread on plate B.
   D. Finally, divide the number of colonies on plate B by the mass of *p*KAN. Express the colonies/µg *p*KAN in standard exponential notation.

4. What factors might influence transformation efficiency?

5. You have introduced *Your Favorite Gene (YFG)* into *p*KAN, and use the recombinant *p*KAN/YFG to transform a culture of *E. coli*. Given:

   A. You begin with 0.2 µg of *p*KAN/YFG
   B. You achieve an efficiency of 106 transformants/µg.
   C. *p*KAN has an average copy number of 100 molecules/cell.
   D. *In log growth, E. coli* replicates an average of one every 20 minutes.

   Assuming that your bacteria enter into log growth 80 minutes following transformation, how many copies of *YFG* would you have 3 hours after transformation?

APPLICATION

How is transformation used in agriculture?

What would be a specific agricultural application of recombination?

GLOSSARY OF TERMS

bacteriophage: A virus that infects a bacterial cell; also called a phage.

competent cells: Bacterial cells that have been treated with cold calcium chloride solution followed by a heat shock; this procedure increases the ability of the cell to take up plasmid DNA increasing the rate of transformation.

DNA ligase: An enzyme that catalyzes the joining of two DNA molecules.

endonuclease: (*endo*—within; *nude*—nucleus; *ase*—enzyme) An enzyme that targets specific base-pair sequences in a strand of DNA or RNA, which produces shorter lengths of the nucleic acid.

plasmid: A small circular piece of DNA that is separate from the bacterial chromosome; carries accessory (additional functional) genes. Plasmids range in size between 1,000 and 100,000 basepairs and may replicate (duplicate) independently of the bacterial chromosomes.

sterile technique: Also referred to as *aseptic technique,* the use of sterile equipment and sterilized surfaces in an attempt to reduce the rate of contamination of a specific microbial culture by unwanted bacterial and fungal cells.

transformation: The assimilation of external genetic material by a cell.

vector: In molecular science, a vector is a DNA molecule that allows foreign DNA to be incorporated, transferred into, and replicated within a host cell.

EXERCISE 10

Plant Reproduction

OBJECTIVES

Upon completion of this lab exercise, you should be able to:

- distinguish between vascular and nonvascular plants and give an example of each type of plant.
- recognize the male and female cones of a coniferous gymnosperm.
- draw and label the stages in the life cycle of a pine.
- define angiosperm, and describe the sexual reproductive cycle of angiosperm.
- identify the male and female gametes of an angiosperm.
- relate the shape of the flower to the method of pollination.
- describe the various mechanisms of pollination found in angiosperms and describe the economic/agricultural implications.
- relate seed shape to seed dispersal.
- list some mechanisms for the asexual propagation of flowering plants.

Suggested Reading:

Chapters 6, 7, and 8 of *The Science of Agriculture: A Biological Approach*, 4th edition.

INTRODUCTION

Plants play an extremely important role in the balance of nature. Plants modify both local and global climate patterns by adding water vapor into the atmosphere through a process called **transpiration**. Plants also release oxygen (O_2) into the atmosphere. The oxygen is a by-product of **photosynthesis**, and oxygen is required by all animals to sustain life. Additionally, plants provide food for most animals, some bacteria, and some fungi. Humans have cultivated plants for several centuries to produce food. Studying the reproductive cycles of plants allows the commercial and home producer to maximize crop yield by understanding pollination mechanisms, fruit set, and seed production.

Members of the plant kingdom can be divided into two groups, **vascular** and **nonvascular** plants. Vascular plants have specialized conductive tissues that transport water and nutrients through the plant. Ferns, pine trees, tomato, cotton, and tobacco are all examples of vascular plants. Ferns have **vascular tissue** only in the stems and leaves. The rootlike structures lacking vascular tissue are called **rhizoids**. Pine trees have "true" roots, stems, and leaves, all possessing vascular tissue. In contrast, the simplest members of the plant kingdom, like bryophytes (mosses and liverworts), lack vascular tissue. Water and nutrients are transported between cells by the processes of osmosis and diffusion.

Hypothesis

Write a hypothesis that relates the presence/absence of vascular tissue to plant size. Give examples of plants that fit your hypothesis.

PLANT REPRODUCTION

All plants undergo **alternation of generations** in their life cycles. Alternation of generations is the alternation of a sexually reproducing stage, the **gametophyte**, with an asexually reproducing phase, the **sporophyte**. In more primitive organisms such as algae, these alternate forms may exist independently. In some marine algae and fungi, the sporophyte and gametophyte look very different. There were several instances in which the sporophyte and gametophyte were thought to be different species. When scientists learned to culture these algae in the lab, they were surprised to find that these very different-looking organisms were actually different phases in the life cycle of a single organism! In mosses, the sporophyte is found growing out of the female reproductive structure at the top of the female gametophyte plant. In more advanced land plants such as pines and flowering plants **(angiosperms)**, the sporophyte is the predominant form. When you look at a pine tree or a tomato plant, you are looking at the sporophyte. The sexually reproducing gametophyte has been reduced to a few cells found inside the male and female reproductive structures. In a pine tree, the female gametophyte is found inside an immature seed cone. Similarly, the male pine gametophyte is found within the immature pollen cone. This pattern holds for flowering plants as well. The female gametophyte is located inside the ovary (female portion) of the flower, and the male gametophyte is found within the stamen (male portion) of the flower.

In seed plants, the male gametophyte produces **gametes** (sperm) that are housed in a special delivery system, the pollen grain. The shape of pollen is highly variable. Pine pollen has two wings, closely resembling the shape of a Mickey Mouse hat. Lily pollen is football-shaped with sharp spines protruding from the surface. The shape of the pollen grain provides clues about the pollination mechanism. Now you will examine an angiosperm flower to learn the male and female parts. It is necessary that you understand the structure of a flower before learning how reproduction in flowering plants occurs.

There were several life cycle modifications among the seed plants that have allowed these organisms to effectively exploit terrestrial habitats that are inhospitable to algae, nonvascular plants, and primitive vascular plants. These modifications include (1) the reduction of the gametophyte to several cells protected by the sporophyte reproductive tissues; (2) pollination, eliminating the dependence on water for successful reproduction; and (3) a seed containing stored food reserves and a protective seed coat.

CONIFERS (DIVISION CONIFEROPHYTA)

Members of the division Coniferophyta are abundant; some 550 species have been described. Conifers are **gymnosperms**, plants that produce seeds lacking the protective integuments of angiosperm seeds. The Coniferophyta are found in regions of the world in which length of the growing season is shortened due to latitude or longitude. Most members of the Coniferophyta are evergreen, but a few deciduous members can be found in semitropical and tropical areas of the world. Permanent leaves allow conifers to take advantage of any sunlight available for photosynthetic activity. The shape of the needles and growth habits of the northern and southern temperate latitude trees allow conifers to withstand large snow loads without losing branches. Conifers are economically important; many of the trees that supply lumber and pulp are conifers: pine, hemlock, cypress, fir, and spruce.

1. Male gametophyte—*Pinus*

   The male gametophyte is produced in a **staminate cone** borne on the terminal portion of branches (see Figure 10-1). **Meiosis** occurs to produce microspores within each segment (scale) of the male cone. The microspores of different conifer species vary in size and shape, modified to enhance pollination in different habitats. Virginia pine (*Pinus virginiana*) has pollen grains that look like Mickey Mouse hats (see Figure 10-2). Texas pine (*Pinus taeda*) has pollen grains with air bladders.

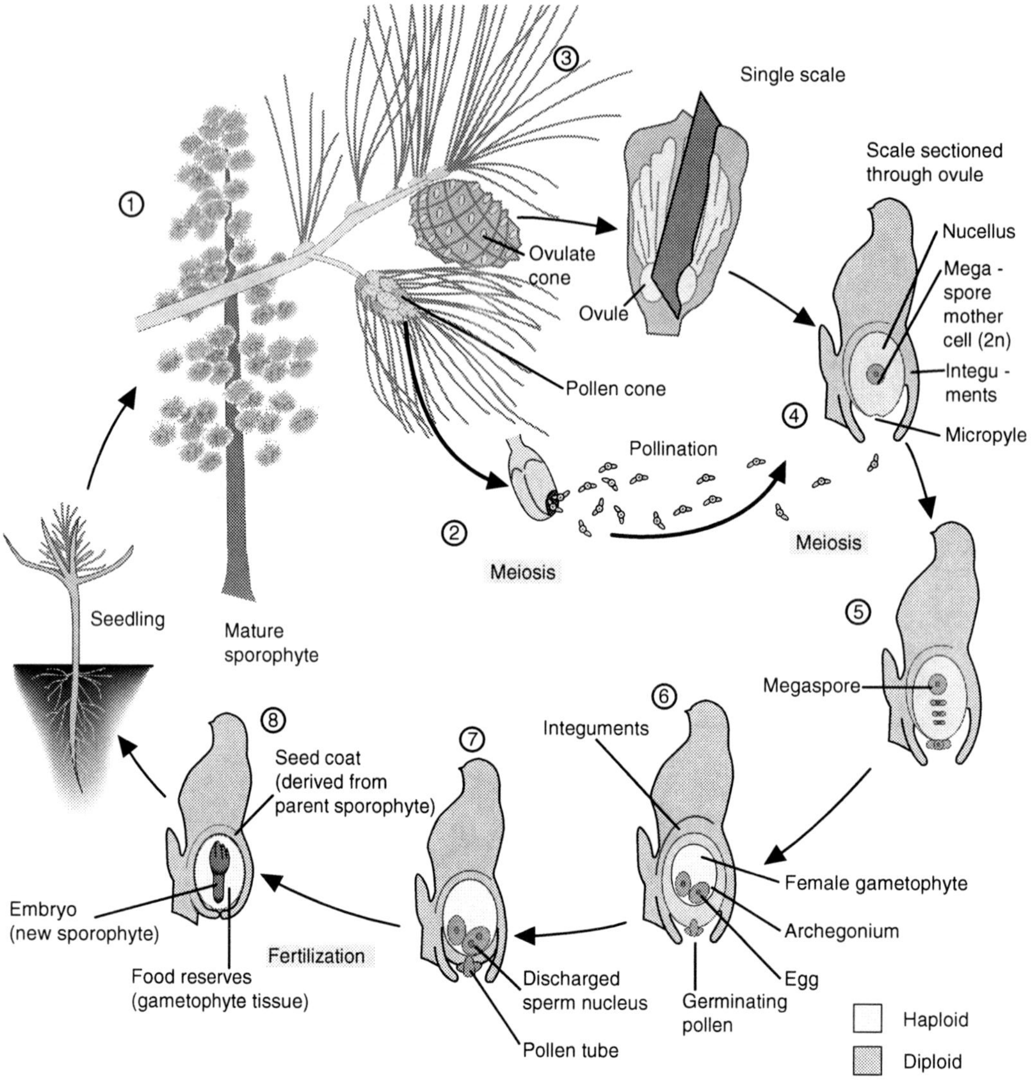

Figure 10-1. Life cycle of a pine.

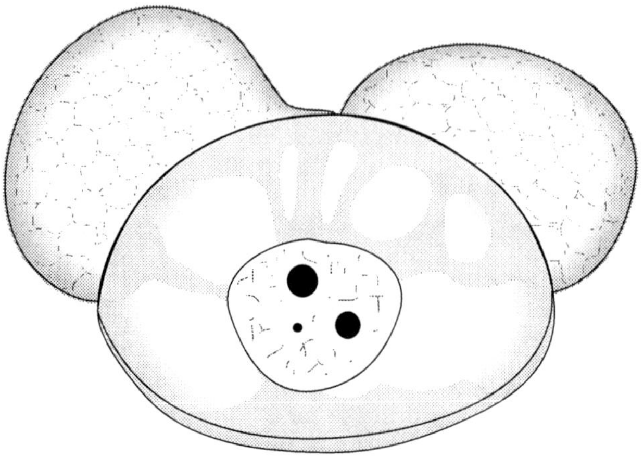

Figure 10-2. Virginia pine pollen.

Hypothesis

Write a hypothesis that relates the shape of conifer pollen to the method of pollen transport. (*Hint:* Think about the shape.)

Why do you think that conifer pollination occurs in the spring instead of the fall or winter? (*Hint:* Think about the typical climate of each season.)

Obtain a preserved staminate cone and remove a scale from the side of the male cone. Mount the scale in an alcohol/glycerol solution on a slide. Crush the scale and apply a cover slip. Examine the slide using the 10×, 40×, and oil-immersion objectives. (*Note:* Use the oil-immersion lens only if the preparation is thin enough to accommodate the lens.) Draw and label what you observe in the space provided.

What is the purpose of the "wings" on the *Pinus* pollen? How is the shape of the pollen grain related to reproductive success?

2. Female gametophyte—*Pinus*

   The female gametophyte is produced in the **ovulate cone** at the apex of lateral branches. The female ovulate cones are produced on the same tree as the male staminate cones; *Pinus* is **monoecious**. Large scales are arranged in whorls around a central axis. The ovule is at the base of the scale (see Figure 10-1).

Obtain a preserved ovulate cone and remove one of the small young ovulate scales at the top of the cone. Place the scale in a depression slide in a solution of water and place a cover slip on the slide. Using the 10× objective lens, draw and label what you see in the following space.

What is the purpose of the scales and bracts of the female cone?

Why are the staminate cones borne on the ends of terminal branches and the ovulate cones on lateral branches?

3. Seeds and seed *dispersal—Pinus*

   The seeds of the pine are released from the female cone after one to two seasons of development, depending on the species. Seeds of the pine are released as the cone desiccates (see Figures 10-1 and 10-3). Some seeds of western conifer species are released only after a fire. What advantage would fire give a seedling?

What role could controlled burning play in the management of commercial conifer stands?

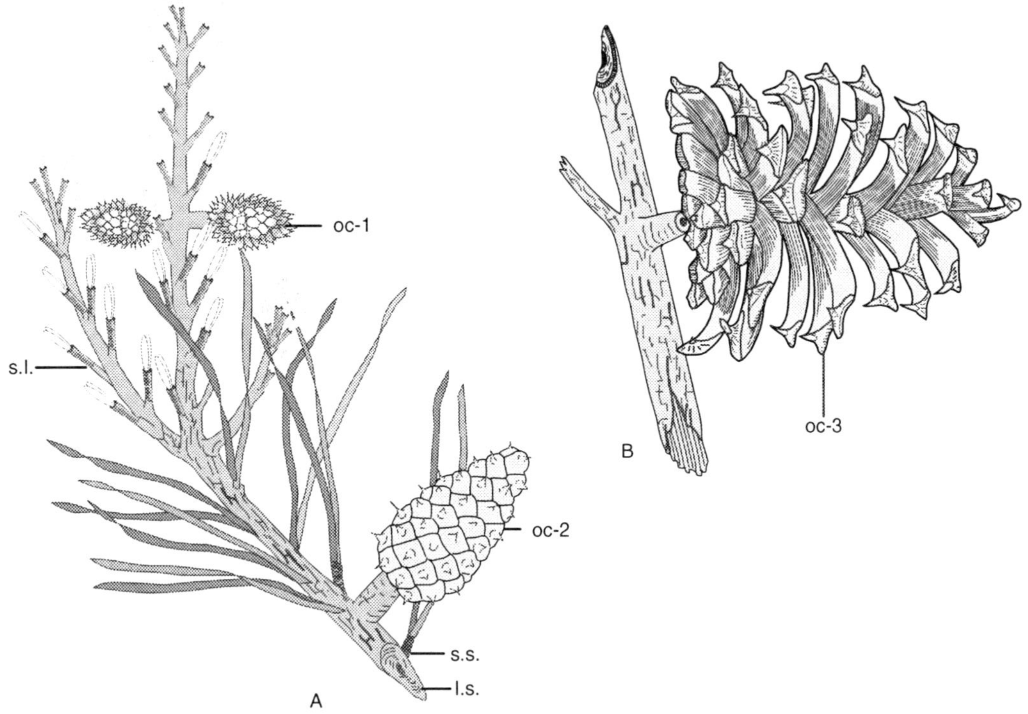

Pinus virginiana Miller. (A) Long shoot with ovulate cone in spring, just after pollination. (B) Ovulate cone shedding seed. l.s., long shoot; oc-1, ovulate cone of the season, just after pollination; oc-2, ovulate cone pollinated 1 year earlier; oc-3, ovulate cone pollinated 2 years earler; s.l., scale leaf; s.s., spur shoot. x1.

Figure 10-3. One, two, and three season ovulate cones of the Virginia pine.

Observe the various pine cones on display. Note the diversity of shape and size among cones of various conifers. To observe the effects of water loss and hydration on a mature ovulate cone, a desiccated cone is placed in a pan of water at the beginning of the lab period. Using a metric ruler, measure the size of openings between scales. Record your measurements and the initial observations on the number in the table below.

Observation of Pine Cone

	Initial (dry)	Final (wet)
Number of open scales	________	________
Average distance between scales	________	________

After an hour, how has the shape of the cone changed?

How might this be related to climate?

Which season(s) of the year do you think pine seeds are most likely to be released?

Observe the pine seeds on display. Draw the seed shape in the space provided.

Given the shape of the seed, what is the most probable mechanism for seed dispersal?

ANGIOSPERMS (DIVISION ANTHOPHYTA)

The angiosperms, members of the division Anthophyta, are the most abundant and widely distributed division of the plant kingdom. There are 235,000 species of flowering plants including species that are important economically, aesthetically, and biologically. These plants have become extremely diverse, with adaptations to a wide range of environmental factors. Possible reasons for their success in comparison with other vascular plants include:

1. Coevolution of the flower with insects (and other animals) as pollinators. Animal pollinators compete with each other for plant nectar, while plants compete for the pollinators. Consequently, through natural selection, a tremendous diversity in flower form, color, and scent as well as plant habitat requirements result in response to the sensory perception of different pollinators and their occurrence in various environments.

2. The evolution of fruit and seed modifications. These provide for efficient seed dispersal by mechanisms such as wind, water, animals, or mechanical means (e.g., expulsion from fruit by pressure). Such modifications have permitted plants to become widely distributed. You will observe a variety of dispersal mechanisms in this exercise.
3. Secondary plant compounds. Many angiosperms have evolved various substances, such as alkaloids, that they produce in complex metabolic pathways. These substances help deter some herbivores from consuming the plants. Humans often use such compounds for medicinal purposes.

FLOWER AND FRUIT STRUCTURE

Angiosperm reproduction produces a flower and a fruit. The flower is produced by the sporophyte plant and is the structure within which **spore** and gamete production occur. The following structures are present within the flower (see Figure 10-4):

receptacle the part of the flower stalk that bears the floral organs

sepals an outermost whorl of flower parts that usually enclose the other flower parts when in the bud (some monocots possess colorful tepals instead of sepals)

petals a whorl of flower parts, usually colored and inside the sepals

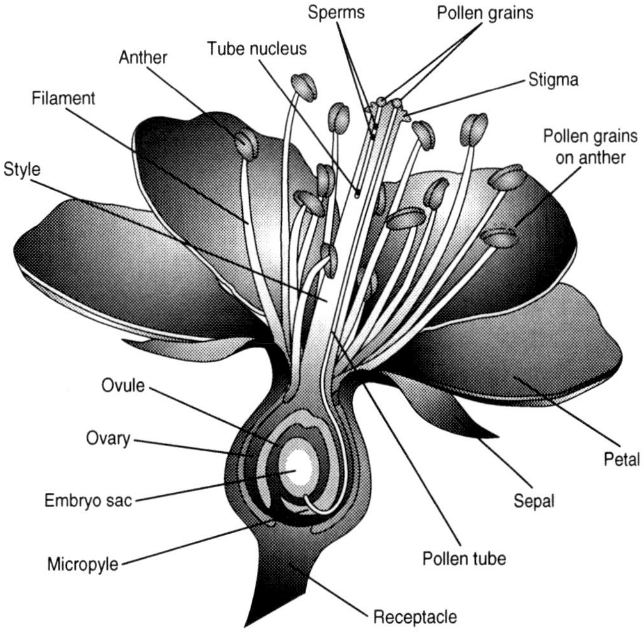

Figure 10-4. Parts of a flower.

anthers the pollen-bearing portion of the flower, on the upper portion of the stamen
filament the stalk of the stamen supporting the anther
stamen collectively, the filament and the anther (sometimes referred to as the *androecium*)
pollen the male gametophyte of seed plants
ovary an enlarged basal portion of a flower connected to the stigma and style within which ovules are produced, becoming the fruit
style slender column of tissue which arises from the top of the ovary and through which the **pollen tube** grows
stigma the receptive surface on the style that receives the pollen and on which the pollen germinates
pistil collectively, the stigma, style, and ovary (sometimes referred to as the *gynoecium*)
ovules structure in the ovary that contains the megaspore that develops into the embryo sac (the female gametophyte), plus outer layers of sporophytic tissue (the *integuments*); becomes the seed following fertilization

Not all flowers are complete, and the inclusion of these structures varies considerably among flowers. Sometimes these parts are fused, missing, or highly modified. Some flowers are monoecious, so the plant has separate male and female flowers. Other plant species are dioecious, with separate male and female plants. American hollies and some grapes are examples of dioecious plants. In the cultivation of dioecious species, it is necessary to plant a "pollinator" (male plant) in order to produce fruit.

There are two classes (types) of angiosperms, the *monocots* and the *dicots*. The two classes vary in many structural features. In the monocot flower, the parts such as sepals, petals, stamen, ovaries, or sections, etc., occur in threes. In the dicot, such structures occur in fours or fives or multiples of these (see Figure 10-5).

From a biological perspective, the function of the flower is to produce gametes and produce a fruit with seeds, following pollination and fertilization. Fruits are mature ovaries containing seeds. Fruits function to protect the seeds during maturation and are modified in many ways for seed dispersal. Each **seed** includes an embryo with a food reserve and a seed coat. There are structural differences in monocot and dicot seeds. The most obvious difference is that the embryo of a dicot seed has two large food storage organs, **cotyledons**, whereas the embryo of a monocot has only one cotyledon.

The sporophyte is independent and conspicuous in angiosperms. It may occur in such diverse forms as an herbaceous plant, a tree, a shrub, or a vine for example. The gametophyte is reduced to a few cells embedded within the flower. Gamete production in angiosperms is unique; the process of **double fertilization** follows pollination in flowering plants (see Figure 10-6).

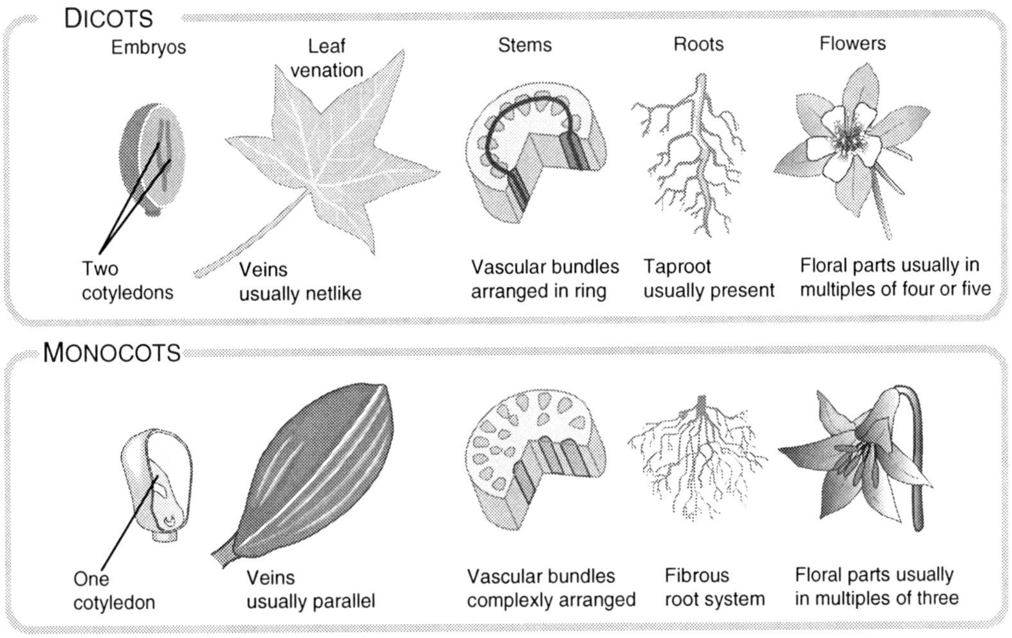

Figure 10-5. A comparison of monocot and dicot characteristics.

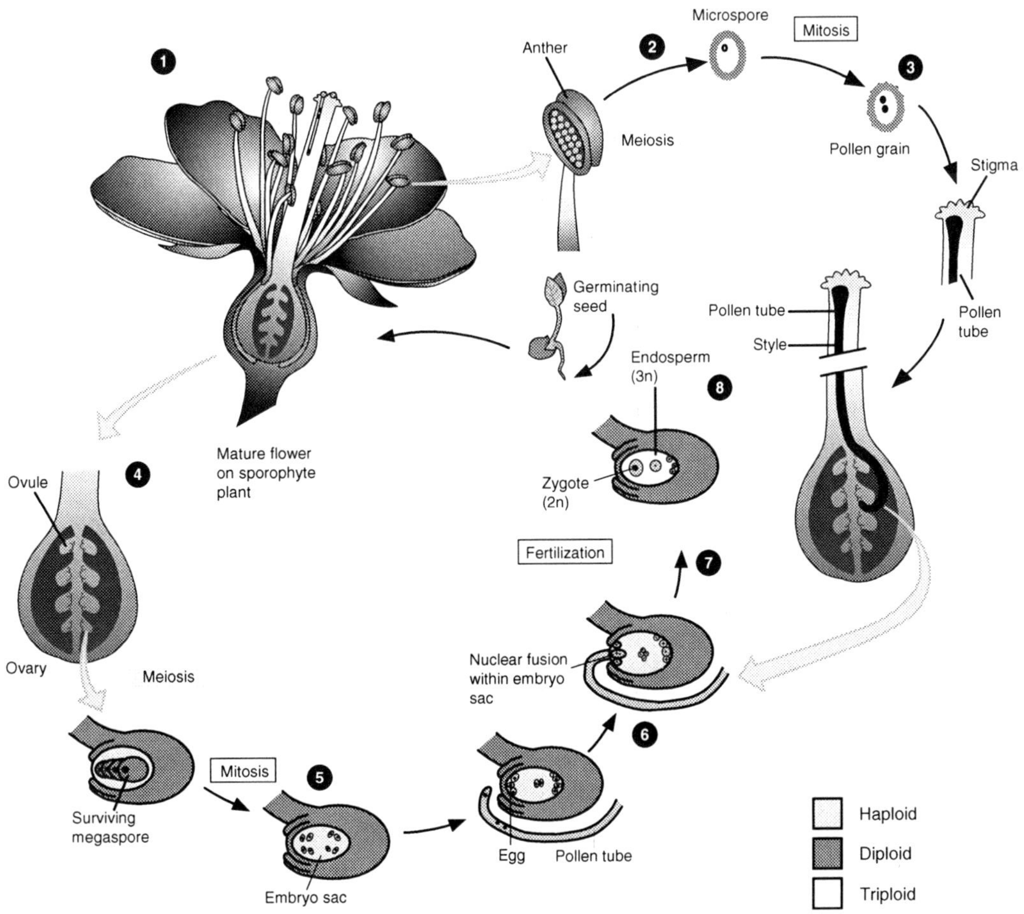

Figure 10-6. Life cycle of a flowering plant.

1. In the anther, microspore "mother" cells develop and each divides by meiosis to produce four microspores. Each microspore then divides to form a two-celled male gametophyte, the pollen grain. The pollen grain divides again, producing a tube nucleus and two sperm cells. The male gametes are sperm cells.
2. Development of the female gametophyte occurs in the ovule, and produces the female spore cell. Following meiosis to produce four megaspores, only one of the megaspores undergoes three rounds of **mitosis** to produce a seven-celled, eight-nucleate structure, which is the female gametophyte, or embryo sac.
3. Pollination occurs with the transfer of pollen from the anther to the stigma.
4. The pollen grain germinates and a pollen tube grows through the stigma and style to the ovule in the ovary.
5. **Double fertilization** occurs: (1) one of the pollen grain sperm nuclei fuses with the egg to form a **diploid** (2N) **zygote**; and (2) the other sperm nucleus fuses with two polar nuclei to form triploid (3N) **endosperm**.
6. The seed coat of the seed forms from the integument, derived from the sporophyte tissue of ovary that surrounds the ovule.
7. The fruit develops from the ovary. The seed matures and is released from the fruit as an embryonic sporophyte.
8. The seed may remain dormant until conditions are favorable for a particular species to germinate.

FLOWER STRUCTURE

1. Working in groups of two, obtain a fresh monocot flower (a gladiolus) and a dicot flower (a petunia, stock, or tobacco) for dissection.
2. First, examine the monocot flower. Note that the two leaflike structures at the base of this flower are not sepals but are actually modified leaves called bracts. The sepals are the three outer petal-like appendages, which are called tepals when they look like petals, as in this flower.
3. Slice lengthwise through the tepals and ovary with a razor blade, so that the reproductive structures can be viewed in a longitudinal section. Count the floral parts and draw the monocot flower, identifying the parts using Figure 10-4.
4. Make a cross section (across the long axis) through the ovary of the gladiolus and examine it under the dissecting microscope. Draw the section as viewed and label the ovules.

Cross section of monocot flower ovary

5. Follow the same procedure to dissect the dicot flower.

   Cross section of dicot flower ovary

Hummingbirds frequently pollinate flowers that are bright red or orange with tubular shapes. Many plants that are pollinated by bees are bright yellow, blue, or purple and have strong, sweet scents. If these flowers are examined under ultraviolet light, markings that guide the bee to the anthers are evident. Some flowers have evolved unique odors to attract pollinators. A cactus pollinated by carrion beetles produces a flower that smells like rotting flesh. Beetles and flies living in rotting flesh or dung pollinate flowers with obnoxious odors. Some flowers are pollinated at night by nocturnal bats and moths. For example, the yucca plant is pollinated nightly by a yucca moth.

Observe the flowers on demonstration and suggest possible pollinator.

daylily	________	red salvia	________
buttercup	________	night blooming cactus	________
orchid	________	rose	________

Your teacher may show you some slides of other flowers, some of which may be of commercial importance. Write the names of the plants and their pollinators in the space provided.

POLLEN TUBE

1. Apply a film of pollination gel to the center of a slide. Close gel jar.
2. Hold slide under yellow anthers of *Tradescantia* flowers.

   Do not remove the entire anther; follow step 3.
3. Gently tap anthers twice to release pollen. Place slide on black surface to see pollen.
4. Add 1 drop of water. Mix pollen, water, and gel, using a toothpick. The mixture should be lumpy.
5. Drop cover slip onto slide. Don't squash the pollen mixture.

   This may be the only slide you'll ever make where it is advantageous to have air bubbles. Air bubbles provide oxygen during germination.
6. Use moderate light levels on the microscope to make your observations. Do not overheat slide (microscope lights are hot). The pollen will die with prolonged exposure to heat.
7. View using the 10× objective until you see pollen grains ("footballs").
8. Switch to the 40× objective when germination begins, and observe.
9. Reexamine at 15-minute intervals to observe a germinating pollen tube. Turn off microscope light between viewings.
10. Add distilled water to the edge of the cover slip as needed to rehydrate the preparation.
11. Look for the nucleus in the tube when the pollen tube has germinated. Draw the pollen grain, showing the surface texture and the tube.

FRUIT STRUCTURE

1. Obtain an immature okra pod (edible at this stage). Make a cross section through the fruit and draw the developing seeds in the pod. How does the okra pod compare with the gladiolus ovary?

2. Draw a mature okra fruit and indicate how the seeds are released from the pod.

Fruits are often referred to as vegetables, even though this is not botanically correct. The term *vegetable* refers to a vegetative (non-reproductive) part of the plant. Identify the following items as a vegetable (as used botanically) or as a fruit.

green beans	________	lettuce	________
carrots	________	squash	________
turnips	________	peas	________
spinach	________	watermelon	________
okra	________	eggplant	________
white potatoes	________	beets	________

How are the following plants adapted for seed dispersal?

tomato	________	coconut	________
maple	________	milkweed	________
cocklebur	________	cherry	________

Hypothesis

Write a hypothesis that relates seed shape and size to the type of dispersal mechanism employed by the plant. Support your hypothesis with evidence from your "Fruit Structure" observations.

ASEXUAL PROPAGATION

Sexual reproduction in flowering plants produces seeds. The plants that develop when the seeds germinate are genetically different from the parent plant(s). Sexual reproduction increases the genetic variation within a population of plants. Plants are specifically bred for certain desirable characteristics: sweetness in corn, nematode resistance in cotton, fusarium wilt resistance in tomatoes, and mosaic virus resistance in tobacco are examples of highly desirable genetic traits. In such instances, it might be preferable to produce plants that are exact genetic copies of the parent and thus contain all of the desirable traits. Asexual propagation produces new plants genetically identical to the parent plant.

Mode of reproduction	Species	Description
Propagation technique		
Parthenogenesis	Rose, orange	Embryo develops from unfertilized **haploid** egg or from tissue surrounding the embryo sac
Tissue culture	Carrot, corn, wheat, rice, orchid	New plant arises from undifferentiated tissue derived from the parent plant–laboratory technique
Vegetative	African violet, hosta	New plant develops from tissue which is separated from parent plant
Reproduction by modified stems		
Bulb	Onion, tulip	Axillary bud on underground stem gives rise to new bulb
Corn	Gladiolus	Axillary bud on short, thick underground stem produces new plant
Rhizome	Running grasses	New plants develop at nodes on underground stems
Runner	Strawberries	New plants develop from nodes on horizontal surface roots
Tuber	Potato	Enlarged tips of underground rhizomes produce new plants

Table 10-1. Types of asexual propagation.

Look at the display of asexually propagated plants on display. Indicate the type of propagation used next to the name of the plant.

garlic	________	Easter lily	________
Zoysia (grass)	________	lily of the valley	________
crocus	________	chives	________
tobacco	________	orange	________
costal bermuda	________	*Bahia*	________
bugleweed (ground cover)	________	jade plant	__________

Under what conditions do you think that increased genetic diversity of a crop plant would be desirable?

What is the relationship between monoculture growing practices and the need for the development of new and improved strains of crops/plants?

GLOSSARY OF TERMS

alternation of generations: The alternation of an asexually reproducing form (the *sporophyte*) with a sexually reproducing form (the *game-tophyte*).

angiosperm: (*angion*—comb form; *sperma*—seed, sperm) A division (phylum) of the Plant Kingdom which produce flowers, undergo double fertilization, and possess seeds with stored food reserves.

antheridium: (*anthems*—flowery) In mosses and ferns, the antheridium is a multicellular sperm-producing structure surrounded by sterile (nonreproductive) cells. In the algae, the antheridium is a unicellular structure in which sperms are produced.

archegonium: (*arch*—primitive; *gona*—reproduction) A multicellular egg-producing female reproductive structure which is surrounded by sterile (nonreproductive) cells.

cotyledon: (kotytē—cup-shaped, hollow) The seed leaves of an angiosperm embryo.

diploid (2N): (*di*—two; *ploid*—onefold) A cell or cells which is formed by the fusion of male and female sex cells, characteristic of the sporophyte generation in most plants.

double fertilization: Unique to angiosperms, the process of fertilization involves the fusion of one sperm nucleus with the egg nucleus (in ovule) forming a diploid (2N) zygote. A second sperm nucleus fuses with two polar nuclei in the ovule forming triploid (3N) endosperm.

endosperm: (*endo*—inside, within; *sperma*—seed) Tissue which nourishes the developing embryo in angiosperm seeds; formed as a result of the union of the two polar nuclei in the ovule and one sperm nucleus derived from the pollen tube.

gamete: (*gamos*—marriage) Sex cells which may be the product of meiotic cell division; gametes are haploid (N).

gametophyte: (*gameto*—sex; *phyta*—plant) The gamete-producing phase of plant which undergoes alternation of generations.

gymnosperm: (*gymmnos*—naked; *sperma*—seed) A vascular plant which produces seeds not encased within an ovary.

haploid (N): (*hapl*—simple; *ploid*—onefold) A cell which is the product of meiotic (reductive) cell division in which each chromosome is represented once; characteristic of the gametophyte generation in most plants.

meiosis: (*meiōun*—diminish) A type of cell division in which the chromosome number is reduced by one-half, and one of each pair of homologous chromosomes passes to each daughter cell; usually occurs in gamete-producing cells.

mitosis: (*mitōs*—thread) A type of cell division in which each daughter cell is genetically identical to the parent cell; the chromosome number remains constant in mitotic cell division.

monoecious: (*mon*—one; *oikos*—house) The condition of having male and female reproductive structures within the same organism; hermaphroditic.

nonvascular plant: (*non*—lacking, without; *vasculum*—small vessel) A plant which lacks vascular conductive tissue. Mosses and liverworts are examples of nonvascular plants.

ovulate cone: (*ovulum*—egg) The conifer structure in which the female gametophyte develops. Following fertilization of the egg in the ovule, seeds develop within the cone.

photosynthesis: (*photo*—light; *synthe*—to make) A light-requiring metabolic pathway in which chlorophyll-containing plants and protists manufacture carbohydrates from carbon dioxide (CO_2) and water (H_2O).

Oxygen given off as a by-product of this reaction. Photosynthesis occurs in the chloroplasts of the cell.

pollen tube: A tube which forms when the pollen grain germinates on the surface of the stigma; grows down through the style and delivers the sperm nuclei to the ovule.

rhizoid: (*rhiza*—root) Rootlike structures lacking vascular tissue.

seed: A complex organ consisting of a seed coat, embryo, and a food reserve which develops as a result of double fertilization in angiosperm flowers; germinates to develop into a new individual when conditions are favorable for growth.

sporangium: (*spora*—seed or spore; *angeion*—vessel) A structure in which asexual spores are produced.

spore: (*spora*—seed or spore) An asexual reproductive cell which can develop into a new individual without fusing with another cell which contains little or no stored food reserves.

sporophyte: (*sporo*—spore-producing; *phyta*—plant) The spore-producing phase of a plant which undergoes alternation of generations.

staminate cone: (*stēmōn*—thread) The male cone of a conifer in which the male gametophyte develops; pollen is released from the staminate cone.

transpiration: The process of water evaporating from the surface of a plant.

vascular plant: (*vasculum*—smail vessel) A plant which has vascular tissue, xylem and phloem, for the conduction of water, nutrients, and minerals in the plant. Ferns, conifers, and angiosperms are examples of vascular plants.

vascular tissue: (*vasculum*—smail vessel) The conductive tissues of plants which conduct food (*phloem*) and water (*xylem*) throughout the body of the plant.

zygote: (*zyg*—yolk) A diploid cell resulting from the fusion of two gametes, develops into a new individual.

EXERCISE 11

The Role of Minerals in Plant Growth

OBJECTIVES

Upon completion of this lab exercise, you should be able to:

- define essential element.
- define mineral nutrient.
- distinguish between:
  - a macroelement and a microelement.
  - a macroelement and a macronutrient.
  - a microelement and micronutrient.
- describe the role of the following essential elements in normal plant functioning:
  - calcium (Ca).
  - magnesium (Mg).
  - nitrogen (N).
  - phosphorus (P).
  - potassium (K).
  - sulfur (S).
- list the deficiency symptoms exhibited by annuals for each of the essential elements.
- construct and test a hypothesis relating to the quality and quantity of growth in a nutrient solution lacking an essential element.

Suggested Reading:
Chapters 8 and 9 of *The Science of Agriculture: A Biological Approach*, 4th edition.

INTRODUCTION

Since the earliest days of agricultural practices, farmers have realized that plants extract substances from the soil. All plants require certain chemical **elements** to grow and complete their reproductive cycles. Elements that are necessary for normal growth and reproduction are called **essential elements**. In the absence of an essential element, plants will develop symptoms of a nutrient deficiency. The symptoms will disappear with the addition of the missing element. The addition of natural substances such as manures and composts may increase crop productivity because they provide the nutrients essential for growth. It was not until early in the 20th century that scientists identified all 16 essential elements (see Table 11-1). A **mineral nutrient** is an **inorganic ion** (noncarbon-containing charged particle) that is taken

Element	Chemical Symbol
Macroelements	
Carbon	C
Hydrogen	H
Oxygen	O
Nitrogen	N
Phosphorus	P
Sulfur	S
Potassium	K
Calcium	Ca
Magnesium	Mg
Microelements	
Boron	B
Chlorine	Cl
Copper	Cu
Iron	Fe
Manganese	Mn
Molybdenum	Mo
Zinc	Zn

Table 11-1. Essential elements required for the growth of vascular plants.

up by plant roots from the soil. Mineral nutrients are required for normal plant growth. An example of a mineral nutrient is nitrate (NO_3^-). Fertilizer is usually labeled with three numbers, such as 20-19-17, which represent three mineral nutrients necessary for plant growth. The first number represents the percentage of total nitrogen, usually in the form of nitrate. The second number, 19, represents the percentage of available phosphorus, in the form of phosphoric acid. The third number, 17, is the percentage of water-soluble potassium (K_2O).

Macroelements (see Table 11-1) are essential elements that are required in relatively large amounts for normal growth and reproduction. Plants use carbon (C), hydrogen (H), and oxygen (O) in the greatest quantities needed for normal metabolic activities. Carbohydrates and lipids are composed of carbon, hydrogen, and oxygen. Nucleic acids and proteins also contain large amounts of carbon, hydrogen, and oxygen. There are six mineral macroelements: nitrogen (N), phosphorus (P), sulfur (S), potassium (K), calcium (Ca), and magnesium (Mg). **Microelements** (see Table 11-1) are required in comparatively smaller quantities or normal plant growth. Boron (B), chlorine (Cl), copper (Cu), iron (Fe), manganese (Mn), molybdenum (Mo), and zinc (Zn) are the essential microelements. Microelements are also referred to as *trace* or *minor* elements.

Plants may not always be able to extract and metabolize the essential elements from the soil in elemental form. In some cases, the element in pure form may be toxic or chemically unstable. Plants frequently utilize essential elements in an alternative form; the essential element is combined with other elements to form a mineral nutrient. **Macronutrients** are the chemical form of macroelements, excluding carbon, hydrogen, and oxygen, used by the plant. Nitrogen (N) is utilized as nitrate (NO_3^{-2}) or as the ammonium ion (NH_4^+, under special conditions). Similarly, **micronutrients** are the usable form of microelements. By way of example, iron may be absorbed as one of two ions: ferrous (Fe^{+2}) or ferric (Fe^{+3}).

All of the essential mineral nutrients have very specific functions in the maintenance of normal plant growth. The chlorophyll molecule contains a central magnesium atom. Chlorophyll is required for photosynthesis to proceed, and a magnesium deficiency results in decreased photosynthetic activity and a yellowing of the plant (chlorosis). Nitrogen is required for the formation of proteins and nucleic acids (DNA and RNA). A nitrogen deficiency severely limits plant growth. Phosphorus is found in adenosine triphosphate (ATP), the energy carrier molecule for all cellular activities, and proteins and nucleic acids (DNA and RNA). Since ATP is required for some of the steps in photosynthesis, a lack of phosphorus limits the plant's ability to produce the sugars essential for normal plant cell functioning. Potassium is involved in the process that regulates the opening and closing of plant stomata. The stomata must open in order to admit carbon dioxide (CO_2) into the leaf. Oxygen (O_2) exits the leaf through the stomata. If the stomata fail to open, photosynthesis cannot occur because carbon dioxide is one of the required starting materials for this process. Plants with potassium deficiencies may have leaves that appear green and healthy. In comparison with magnesium deficient plants, potassium deficient plants have greatly decreased levels of photosynthetic activity (Moss, 1984). The negative impact of potassium deficiency on crop yield is significant. Calcium is required for the formation of plant cell walls, and sulfur is essential in the construction of disulfide bonds in protein molecules.

Many factors influence the uptake of soil mineral nutrients. One of the most significant is soil pH. The uptake of many metal ions such as iron, aluminum, copper, and zinc increase as soil pH decreases (see Figure 11-1). In the extremely acidic clay strip mine soils in the southeastern United States, plants may exhibit symptoms of iron or aluminum toxicity owing to the low soil pH. Conversely, crops grown in alkaline soils (pH $\geq$ 8) may exhibit symptoms of iron deficiency when an analysis of the soil reveals that there is sufficient available iron. Under alkaline conditions, the uptake of metal **cations** is reduced. Producers must add soil amendments, which bring the soil pH closer to 6.5–7.5 (see Figure 11-2).

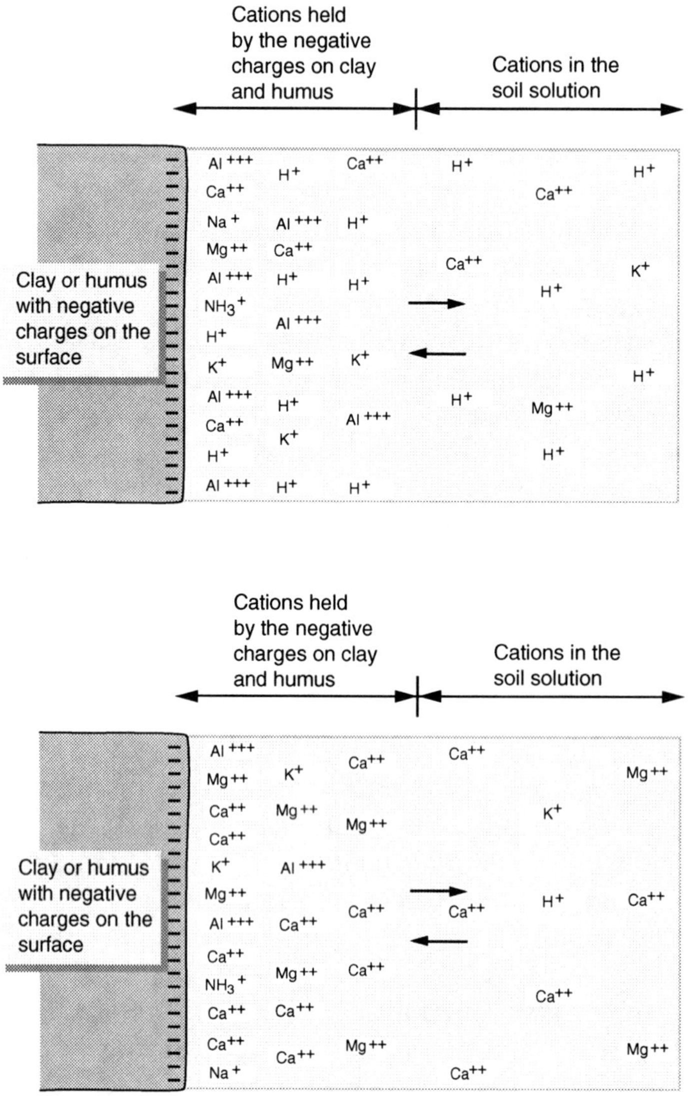

Figure 11-1. Relationship between cation (positively charged ion) exchange and soil pH. Upper figure represents acidic soil-excess soil aluminum ions are a result of H^+ reaction with $Al(OH)^3$. The lower figure shows the reduction in aluminum ions with an increase in soil pH to 6.5.

$$\boxed{\text{Micelle}}\begin{matrix} H^+ \\ \\ H^+ \end{matrix} \quad + \quad Ca^{++} \longrightarrow \boxed{\text{Micelle}}\; Ca^{++} \; + \; 2H^+$$

Figure 11-2. Response of a clay micelle to addition of ground limestone (calcium carbonate) to increase soil pH.

MINERAL DEFICIENCIES

Your instructor will provide you with young tomato plants of a uniform age. Each plant should have at least eight true leaves. Gently shake off as much soil as possible and transplant the tomatoes into the pots provided using sterile vermiculite as a potting medium. Your instructor will divide your class into groups of three or four students. Each group of students will have eight plants: Control, Ca-calcium, N—nitrogen, K—potassium, P—phosphorus, Mg—magnesium, S—sulfur, and Fe—iron. You will water your plants daily with one of eight nutrient solutions. The control solution has all of the necessary macronutrients (Hoagland's solution; Hoagland and Arnon, 1938; Moore, 1973). A test solution missing one of the essential macronutrients will be applied as needed to one of the tomato plants in each of the seven treatment groups. You will be given seven deficient solutions, each missing one essential macronutrient. Using a permanent marking pen, write the names of the people in your group, the date, and the treatment type (Control, Ca, N, K, P, Mg, S, or Fe) on the plant label provided. When you are finished, your instructor will provide you with a place for your plants. All of the control plants will be placed in flats with capillary matting, and each of the seven treatment groups of plants will be in the same flat with capillary matting. The capillary mats will allow the plants to continually absorb the mineral solutions. You will make and record observations at regular intervals for 10 weeks. At the end of the experiment, all of the data collected by each group will be pooled. Your instructor will help you to summarize the results of the experiment.

The symptoms produced by plant nutrient deficiencies relate to the role the nutrient plays in cellular metabolism. The visible symptoms of several macronutrient deficiencies are summarized below (Noggle, and Fritz, 1976).

Calcium	Symptoms appear at growth points of the roots and stems, flower and fruit set reduced or lacking; root system poorly developed—may be jellylike; leaves yellow, rolled, and curled.

Iron	Effects of iron deficiency systemic; leaf tissue between the veins white and chlorotic—appearing first in youngest leaves; leaf margins and tips scorched, eventual chlorosis/bleaching of all shoots and leaves.
Magnesium	Brittle leaves frequent; green veins with yellow to white leaf tissue between the veins presenting a mottled appearance; leaf drop may occur.
Nitrogen	Shoots spindly—short and thin; growth in young plants stunted; young leaves yellow in appearance, older leaves will be green then turn yellow; leaf drop common.
Phosphorus	Purple accessory pigments (anthocyanins) produce a blue-purple color in the veins; plants may be dwarfed; young leaves blue-green to purple in color.
Potassium	Dead (necrotic) spots appear first on old leaves; irregular leaf surface (crinkled, wrinkled) between the leaf veins; leaves blue-green to purple in color; leaf margins may become bleached (white) or chlorotic (yellow) in appearance.
Sulfur	Stems may be fine/slight; leaf color changing from green to light green to yellow; color change first appears around leaf veins.

Hypothesis

Which mineral deficient solution do you expect will produce symptoms first? Why?

Write a hypothesis that relates amount of plant growth and/or appearance to each specific macronutrient deficiency.

1. Calcium

2. Nitrogen

3. Potassium

4. Phosphorus

5. Magnesium

6. Sulfur

7. Iron

Do you expect to see symptoms in the control plant? Explain your answer.

DATA COLLECTION

You will need to use a loose-leaf notebook or computer spreadsheet program to organize your data. Your instructor will provide you with data sheets for you to make your observations. You will have 13 sheets for your group. The first sheet will be used to make observations at the time of transplanting (before the test solutions are applied). You will use an additional data sheet each week for 10 weeks to record your weekly observations. One data sheet will be used to summarize your group's findings at the end of 10 weeks, and the last data sheet will be used to record and summarize class data.

When calculating your summary data, you need to compare your final data table (after 10 weeks) with your initial data table. You would record a change in average internode distance, or the distance between two nodes or the buds that form leaves (final to initial). For example, if you determined that your average internode distance for the control plant was 2.5 cm and the final value was computed at 6.5 cm, then the change in average stem internode distance was 6.5–2.5 cm or 4.0 cm.

Names: **Date:**

Treatment	Plant height (cm)	Average stem internode distance (cm)	Number of leaves	Leaf appearance	Number of flowers/ fruit	Stem diameter (cm) at first pair of leaves
Control						
Calcium						
Nitrogen						
Potassium						
Phosphorus						
Magnesium						
Sulfur						
Iron						

Table 11-2. Data table for student group.

Names: **Date:**

Treatment	Change in plant height (cm)	Change in average stem internode distance (cm)	Change in number of leaves (+/−)	Leaf appearance	Number of flowers/fruit	Change in stem diameter (cm) at first pair of leaves
Control						
Calcium						
Nitrogen						
Potassium						
Phosphorus						
Magnesium						
Sulfur						
Iron						

Table 11-3. Summary table for student group.

Names:

Date:

Treatment	Change in plant height (cm)	Change in average stem internode distance (cm)	Change in number of leaves (+/−)	Leaf appearance	Number of flowers/fruit	Change in stem diameter (cm) at first pair of leaves
Control						
Calcium						
Nitrogen						
Potassium						
Phosphorus						
Magnesium						
Sulfur						
Iron						

Table 11-4. Summary table for class data.

Results

Using the weekly data collected by your group, plot the following information using a line graph. If you have used a spreadsheet program to store your data, you can graph your results.

1. Time (number of weeks) with plant height (cm)

   Remember that time 0 is your data from your initial observation. Time is on the horizontal *(X)* axis of the graph and plant height (cm) is on the vertical *(Y)* axis of your graph. Remember to divide the squares on the *Y* axis to cover your full range of values. For example, if your control plant began at 4 cm and grew to 25 cm, then the scale on your *Y* axis should be from 0 to 30. You will need to make a line for each of the seven treatments and one for the control. Use colored pencil/pen and give each line a different color. Provide a legend at the bottom of your graph (Red: Control, Black: Calcium, Blue: Nitrogen, etc.).

   Note: Think back to what you learned about experimental design in Exercise 2. The independent variable is plotted on the *X* axis, and the dependent variable is plotted on the *Y* axis.

2. Time (number of weeks) with average stem internode distance (cm)

   Follow the directions in Step 1 and plot time in weeks on the horizontal *(X)* axis and average stem internode distance (cm) on the vertical *(Y)* axis.

3. Time (number of weeks) with stem diameter at the first pair of leaves (cm)

   Follow the directions in Step 1 and plot time in weeks on the horizontal *(X)* axis and stem diameter at the first pair of leaves (cm) on the vertical *(Y)* axis.

DISCUSSION AND CONCLUSIONS

1. Describe the change in overall appearance of the seven treatment plants over the course of the 10-week experiment.

   Calcium

   Nitrogen

Potassium

Phosphorus

Magnesium

Sulfur

Iron

2. How might these changes be related to the metabolic function of the macronutrient in plant cells?

3. Were your results different from your original predictions?

4. Were your hypotheses supported or refuted by the data your class collected?

Summarize your results in terms of implications for farming practices. How might soil pH affect your fertilizing schedule and/or use of "green" fertilizers?

GLOSSARY OF TERMS

anion: Negatively charged ion; chloride ion (Cl^-), nitrate ion (NO_3^-), and phosphate ion ($H_2PO_4^-$) are all anions.

cation: Positively charged ion; sodium ion (Na^+), potassium ion (K^+), and ferric ion (Fe^{+3}) are all cations.

element: A substance that cannot be broken down into other substances by ordinary chemical and physical means; elements are atoms that have the same number of protons (subatomic particles with a positive charge). *Note:* The number of protons in an element is also called the *atomic number.*

essential element: An element that is required for normal plant growth; in the absence of the element, the plant is unable to complete the vegetative and/or reproductive part of its life cycle.

inorganic: A chemical compound that lacks both carbon (C) and hydrogen (H); for example, carbon dioxide (CO_2) and sodium chloride (NaCl) are inorganic compounds.

ion: A charged particle; an atom or group of atoms which carries an electrical charge; a positive charge results from the loss of one or more electrons (K^+, potassium ion) and a negative charge results from the gain of one or more electrons (NO_3^-, nitrate ion).

macroelement: Elements that are required in large amounts for normal plant growth. There are nine macroelements: carbon (C), hydrogen (H), oxygen (O), potassium (K), calcium (Ca), magnesium (Mg), nitrogen (N), phosphorus (P), and sulfur (S).

macronutrient: Chemical form of a macroelement that is utilized by a plant for normal growth; K^+ (potassium ion), Ca^{+2} (calcium ion), Mg^{+2} (magnesium ion), $N0_3^-$ (nitrate ion), $H_2PO_4^-$ (phosphate ion), and SO_4^{-2} (sulfate ion).

microelement: Elements that are required in relatively small amounts for normal plant growth, may also be called minor or trace elements; boron (B), chlorine (Cl), copper (Cu), iron (Fe), manganese (Mn), molybdenum (Mo), and zinc (Zn) are microelements.

micronutrient: Chemical form of a microelement utilized by a plant for normal growth; boron (B) as boric acid (H_3BO_3), chlorine as the chloride ion (Cl), copper as the cupric ion (Cu^{+2}), iron as the ferrous ion (Fe^{+2}) or ferric ion (Fe^{+3}), manganese as the manganous ion (Mn^{+2}), molybdenum as the molybdate ion (MoO_4^{-2}), and zinc as the zinc ion (Zn^{+2}).

mineral nutrient: Inorganic ion that is required for plant growth and is extracted by the plant's roots from the soil.

organic: A compound that contains both carbon (C) and hydrogen (H) atoms; may also contain other atoms such as oxygen (O), nitrogen (N), phosphorus (P), and sulfur (S), for example; methane (CH_4) and glucose ($C_6H_{12}O_6$) are organic compounds.

EXERCISE 12

Animal Systems/Renal Physiology: What Is Wrong with This Animal?

OBJECTIVES

Upon completion of this lab exercise, you should be able to:

- list the three functions of the excretory system.
- define homeostasis.
- describe the processes involved in the formation of urine.
- list the parts of the nephron involved in urine formation.
- describe the effects of diet on blood and urine pH in horses and cattle.
- relate the effects of abnormal diet components to poor renal function in horses.
- describe the methods used in urinalysis:
  - urine pH.
  - presence/absence of bilirubin.
  - presence/absence of blood: hemoglobin, myoglobin, or erythrocytes.
  - presence/absence of protein.
  - presence/absence of glucose.
  - urine specific gravity.
  - physical characteristics.
  - presence/absence of casts, cells, bacteria.

Suggested Reading:

This lab will draw from a variety of different topics related to the functioning of animals' bodies. Chapter 10 and 13 of *The Science of Agriculture: A Biological Approach*, 4th edition.

INTRODUCTION

The kidneys play an important role in maintaining homeostasis. *Nephrons,* the functional subunit of the kidney, perform three specific functions:

1. Through the removal of a selective amount of water from the blood, the nephrons control blood concentration and blood pH.
2. Nephrons regulate blood pH by the selective removal/reabsorption of blood electrolytes.
3. Nephrons remove toxic substances from the blood.

Homeostasis is the maintenance of stable internal environmental conditions, usually within a "normal" range. The kidneys play an important role in homeostatic balance. The nephrons remove many substances from the blood and return some substances to the blood that is needed for normal functioning. The remaining materials are eliminated from the body as *urine*.

The purpose of this lab exercise is to provide an understanding of the role of the kidneys in urine formation, including the regulation of urine composition. Through an exploration of several case studies based on actual veterinary clients, the relationship between normal kidney functioning and renal abnormalities resulting from disease/injury should become clearer. You will be asked to serve on a team of "veterinarians" to evaluate an animal patient. You and your team will be provided with a description of the symptoms and possible conditions that may produce these symptoms. Each team will also be provided with background information including technical articles, veterinary medical references, and general books from which to draw information. A urine sample from each animal should be tested to assist you in your diagnosis. Your team will determine which urinalysis test(s) may be appropriate, and you will perform all pertinent tests. Your team will then make a determination as to the possible cause of the animal's symptoms and report your findings to your peers.

URINE FORMATION

There are three processes involved in urine formation: **glomerular filtration**, **tubular reabsorption**, and **tubular secretion**. The **glomerulus** is the portion of the nephron (see Figure 12-1) consisting of a ball of small capillaries. The capillary walls are covered with small pores and the external layer of the glomerulus contains filtration slits. Under normal conditions, the pore and filtration slit size allow only the liquid portion of the blood to pass through the glomerulus into the glomerular (Bowman's) capsule. This liquid, the filtrate, lacks the living cells and large molecules present in circulating blood (see Table 12-1 for normal components of blood for cattle and horses). All of the substances present in the blood plasma are also present in the glomerular filtrate. Water, plasma proteins (such as albumin), nitrogenous wastes (urea and creatinine), **electrolytes** (calcium (Ca^{+2}), chloride (Cl^-), sodium (Na^+), bicarbonate (HCO_3^-), potassium (K^+), and glucose are removed from the blood in the filtration process. When the kidneys are functioning normally, the amounts of these substances are greatly reduced in the urine. The processes of tubular reabsorption and tubular secretion are responsible for the selective reabsorption of substances from the filtrate into the blood. The amount of filtrate passing out of

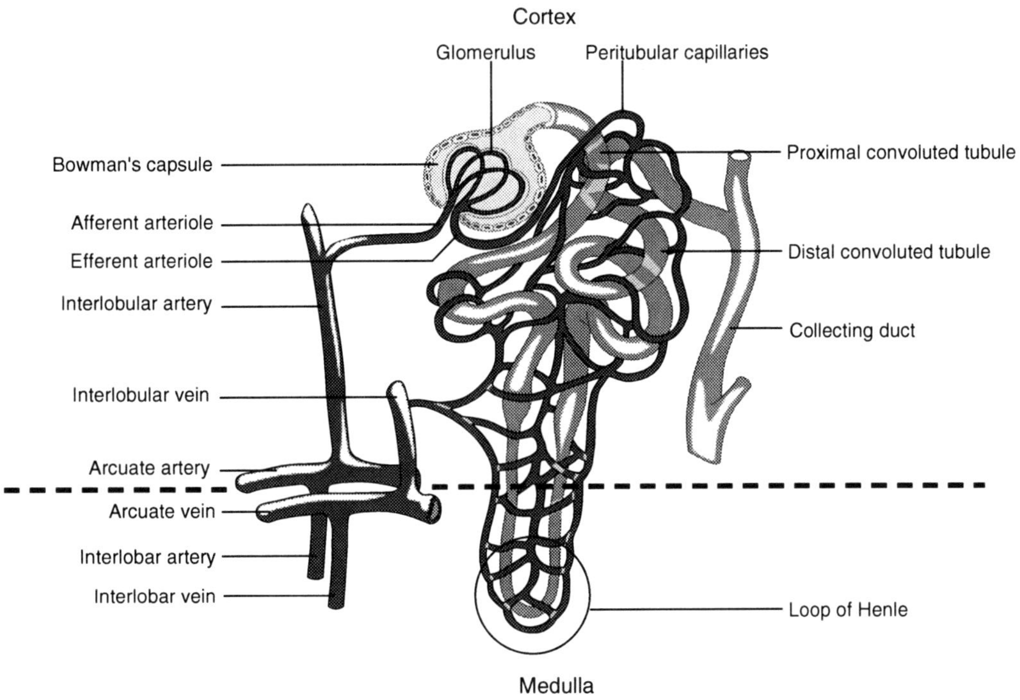

Figure 12-1. Generalized structure of a nephron. The dashed line represents the boundary between the outer cortical (cortex) region and the inner medullary (medulla) region of the kidney.

Animal	Creatinine	BUN*	pH	HCO_3^-	Na^+	K^+	Ca^{+2}	Cl^-
Cow	1–21	6–27	7.31–7.51	20–30	132–152	3.9–5.8	9.12	97–111
Horse	1.2–1.9 (mg/dl)	10–24 (mg/dl)	7.32–7.44	20–28	135–142	2.8–4.0	11.2–13.6	99–105
				Electrolytes measured in mEq/L				

Normal serum bovine values (Divers et al., 1982).
Normal serum equine values (Tennant et al., 1982).

Table 12-1. Normal values for bovine and equine serum biochemistry.

all of the renal corpuscles (glomerulus and capsule) for both kidneys per minute is termed the *glomerular filtration rate* (*GFR*). In livestock, the glomerular filtration rate is estimated using creatine clearance (C_{cr}), which is measured in milliliters per minute per kilogram of body weight (Morris et al., 1984). Filtration rates for commercial livestock vary greatly with the species, age, and condition of the animal (*Note:* normal renal clearance for humans is 125 ml/min). Animals with some renal damage generally have lower values for serum (blood) creatinine clearance (C_{cr}).

The renal tubules (highly coiled thin-walled tubes which are connected to the glomerulus) are the site of reabsorption and secretion. Reabsorption of important substances from the filtrate (such as glucose, Na^+, Cl^-, etc.) occurs in the **proximal convoluted tubule** and the ascending and descending limbs of the nephron loop. Several hormones such as aldosterone and antidiuretic hormone (ADH) are active in this region of the nephron and act to regulate the selective reabsorption of several electrolytes and water. Water is reabsorbed via osmosis in the proximal convoluted tubule and from the **collecting duct** under the regulation of ADH. Tubular secretion is the addition of substances to the filtrate from the blood. Large drug molecules such as penicillin and some ions such as potassium (K^+) and hydrogen (H^+) are removed from capillaries surrounding the **distal convoluted tubule** (peritubular capillaries) and the collecting duct and added to the filtrate at this point. The secretion of excessive hydrogen ions assists in the maintenance of normal blood pH. Abnormal concentrations of substances in the urine can result from the malfunction of a portion of the nephron, abnormalities in the digestive tract, problems with the endocrine system, infection, pregnancy, ingestion of toxins, overmedication, and other physical stresses.

PROCEDURES: URINALYSIS

In the evaluation of any patient, a complete urinalysis includes the following:

1. inspection of the sample to determine general characteristics
2. chemical analysis
3. specific gravity measurements to determine the relative concentration of substances dissolved in the urine
4. examination of any urine sediment using a microscope

PHYSICAL CHARACTERISTICS

The first step in evaluating the animal patient includes measuring the volume of the urine, examining the urine color, clarity, and odor. Each of these observations in combination with the physical symptoms can provide information about the cause of the problem and the overall condition of the animal. Urine volume is measured in milliliters or liters depending on the age and size of the animal and the duration of urine collection (one sample, 1 hour, 24 hours, etc.). Under normal conditions, urine volume is related to water intake. The more water the animal consumes, the more urine the animal will produce and excrete. Urine volume is usually inversely related to urine concentration (as measured by specific gravity). As water intake increases and urine production increases, the lower the concentration of solid substances (called **solutes**) is in the urine. Normal urine color is light yellow or "straw" colored. The color results from the presence of urochrome and urobilin pigments. Dark brown color can result from low water intake or dehydration (as is sometimes observed in lactating animals). Red to reddish-brown urine color may indicate the presence of red blood cells (a condition termed **hematuria**) or the hemoglobin pigment found in red blood cells **(hemoglobinuria)**. The pigment found in muscle tissue, **myoglobin**, can also find its way to the urine **(myoglobinuria)** when muscle tissue damage is excessive, lending a red color to the urine sample. Drugs and vitamins given to animals can also produce abnormally colored urine samples. Since many of the semiquantitative chemical analysis involve a color change on a dipstick or pad, strong color or abnormal color should be noted and recorded as it may interfere with the ability to read the test results. The clarity of the urine sample may suggest infection or injury. For animals other than horses, the urine sample should be clear to somewhat transparent. Horses may produce cloudy or hazy samples as a result of calcium carbonate crystals and/or mucous in the urine. Cloudy samples should be centrifuged and the sediment collected for microscopic examination. Murky samples can indicate the presence of bacteria, red or white blood cells, epithelial tissue cells, sperm (if the animal is male), or some chemical in crystal form. If there is a large quantity of hemoglobin or myoglobin present in the sample, centrifugation will not clear the sample. All frozen or cooled samples should be allowed to come to room temperature before testing. The odor of animal urine varies greatly from species to species and among male and female animals. An animal with a bacterial infection of the urinary tract can have urine with a strong offensive odor. Under rare circumstances, an animal may be concentrating ketones (by-products of fat and protein metabolism) in the urine producing an acetone smell. Diet greatly affects urine odor. Animals that are exclusively carnivorous produce urine with a strong ammonia odor. Herbivore urine may have little or no smell.

There are several tests that are generally performed to examine the chemical composition of urine which involve the use of dipsticks. In some instances, a dipstick test should be followed by more specific assay tests due to the possibility of false negative readings in alkaline (pH greater than 7.5) urine.

pH-Alkalinity and Acidity

pH is inversely related to the concentration of hydrogen ions present in the substance. The higher the concentration of hydrogen ions, the more acidic the substance and the lower the pH value. The pH of urine is most accurately determined with a pH meter, but the pH can be determined with fair accuracy using pH paper. Normal urine pH for equine urine is between 7.0 and 9.0. Bovine urine pH is slightly lower. Excessive protein consumption produces urine with a low pH. Failure to eat can produce urine with a low pH due to the animal's body utilizing its own muscle tissue as an energy source. Excessive exercise or poor exchange of respiratory gases can lower blood pH and, as a result, reduce urine pH. A vegetarian diet produces more alkaline urine. Dip the pH paper in the urine sample. Quickly remove the pH test paper and compare the color of the paper to the indicator scale on the side of the test strip box.

Bilirubin

Bilirubin is formed in the body as a result of the lysis of erythrocytes (red blood cells) and the action of certain cells associated with the immune system (reticuloendothelial cells). Bilirubin circulates in blood plasma in combination with an albumin molecule. (*Note:* Albumins are the largest class of plasma proteins.) Bilirubin combined with albumin is termed *free bilirubin.* Bilirubin is separated from the albumin molecule in the liver where it combines with other liver cell molecules (most notably glucuronic acid or sulfate molecules). Bilirubin in this form is termed *conjugated bilirubin.* When bilirubin is present in the blood plasma in excessive quantities, it may also be present in the urine in large amounts. Elevated urine bilirubin concentrations result in **bilirubinuria**. Urine tests do not necessarily distinguish between free bilirubin and conjugated bilirubin. Excessive amounts of free bilirubin in the urine tend to indicate an excessive level of hemolysis (red blood cell destruction). High levels of conjugated bilirubin are indicative of liver dysfunction. To determine the level of total bilirubin (free bilirubin + conjugated bilirubin) in urine, dip the Bili-Labstix indicator strip in the sample and remove immediately. Compare the strip color to the chart on the side of the bottle.

Heme/Erythrocytes

Hematuria, red blood cells in the urine, is a general indicator of urinary infection, blood vessel damage, inflammation, or irritation (such as a kidney stone). Use a Chemstrip to test for the presence of "heme," the pigment present in red blood cells. Dip the Chemstrip in the sample and remove immediately. Compare the strip to the color chart provided.

Glucose

If glucose is present in the sample, the glucose reacts with an enzyme on the reagent stick. This reaction does not work well at low temperatures, so it is important to allow the urine sample to come to room temperature if the specimen has been refrigerated or frozen. Elevated blood serum glucose levels result in the excretion of the excess glucose by the kidneys (**glucosuria**). Elevated blood sugar (**hyperglycemia**) is usually a product of pancreatic, pituitary, or adrenal **cortex** dysfunction.

Protein

The dipstick test for proteins relies on the use of a pH-sensitive indicator. Alkaline urine may produce a false positive reading. There are many proteins found in the blood plasma that can find their way to urine under abnormal or stressful conditions. Albumin is the most common plasma protein, and it is usually present in urine in trace amounts. However, large amounts of albumin may be excreted in the urine following heavy exercise. **Albuminuria** (excessive albumin levels) can result from high blood pressure, exposure to bacterial toxins or heavy metals, or renal dysfunction. Myoglobin may be found in the urine following heavy exercise or in an anorectic animal. Positive protein dipstick tests should be followed by more sensitive chemical tests to confirm the presence of one or more proteins in the urine.

SPECIALIZED TESTS FOR URINE

Acid Precipitation Test for Protein

Adding a strong acid to the sample can cause protein to precipitate, or fall out of solution. Ten ml of the specimen should be centrifuged at low speed (1,500 to 2,000 RPM) for 5 minutes before the addition of the acid test reagent. Pour off the liquid portion of the sample into a test tube or beaker. Save the solid sediment at the bottom for microscopic examination. Wearing safety goggles and gloves, add a sulfosalicylic acid tablet to the test tube containing the sample (Bumintest, made by Ames), and mix gently. If the solution remains clear, there is no protein present in the urine sample. If the solution is very cloudy, then a large amount of protein is present. In a clinical setting, the cloudiness is assessed on a scale of 0 to 4+.

Glucose

Normal urine contains a trace amount of glucose. Glucosuria, glucose in the urine, can temporarily result from excessive carbohydrate intake. Benedict's reagent is used to test for the presence of glucose in the urine. Obtain a clean test tube and add 2 ml of the urine sample to the tube. Add an additional 2 ml of Benedict's solution to the test tube, and place the test tube in a boiling water bath for 5 to 10 minutes. Remove the test tube from the water bath and compare the color of the tube contents to the scale provided.

Benedict's color	Sugar amount
Blue	None
Greenish-blue	Trace
Green	+ (1)
Yellow	++ (2)
Orange-red	+++ (3)
Red	++++ (4)

URINE SOLUTE CONCENTRATION

Specific Gravity

Specific gravity is the ratio of weight to the volume of a substance compared with a distilled water standard (specific gravity = 1.0). The specific gravity of urine varies with the amount of solids present and normally ranges from 1.001 to 1.035. High specific gravity indicates a high concentration of solutes. These solutes and solids may be blood cells, bacteria, or molecules such as ketone bodies or glucose present in abnormally high amounts. To determine the specific gravity of a urine sample, fill the 25 ml graduated cylinder with the 20 ml of the urine sample. *Do not overfill the cylinder.* Place the urine hydrometer (*urinometef*) in the cylinder containing the urine sample. There is a scale on the side of the urinometer indicating the specific gravity. The specific gravity is equal to the place on the scale where the float rests at the top of the urine column. *Do not throw the sample away—return it to the patient bottle!*

MICROSCOPIC EXAMINATION OF URINARY SEDIMENT

Obtain 10 ml of the urine sample and centrifuge at low speed (1,500 to 2,000 RPM) for 5 minutes. Do not use a higher speed, as it will cause any cells present in the urine sample to fall apart. If you centrifuged a sample earlier for the protein precipitation test, you can use the sediment from the test as long as the sediment did not dry out. Gently mix the remaining urine with the sediment and place a drop of the urine on a slide. Add a cover slip and examine the slide using the lowest power magnification. Adjust the microscope for maximum contrast (dim light, and or diaphragm/condenser closed, condenser down). Large crystals and some larger cells will be visible using the 10× objective. In order to see smaller cells such as red blood cells and bacteria, it is necessary to use the high-power objective lens. There are several different kinds of urine sediments. Casts are precipitated protein formed in the renal tubules. Casts can be made of fat, small protein granules, wax, or in combination with cells. The presence of waxy casts and the larger "broad" casts formed in the collecting ducts of the nephron indicate severe renal damage. Cells from the bladder lining or from the lining of the renal tubules may be found in urine. Bladder lining cells (transitional epithelium) are round to oval in shape. Renal epithelial cells are round and flattened. The presence of these cells indicates severe kidney damage. Leukocytes (white blood cells, WBCs) are present in the urine when an infection is present. The infection is located in the kidney, bladder, ureter, and/or urethra of the animal. Red blood cells are small biconcave disks that have a pale pink appearance. Urine pH, urine concentration, and the temperature of the urine sample influence crystals present in the urine. Many drugs/hormones administered to animals will produce crystals in the urine (see Table 12-2).

Crystal	Shape of crystal	Significance	Urine pH
Ammonium biurate	Yellow to brown spheres with thornlike projections	Normal or liver disease	Neutral-alkaline
Calcium carbonate	Dumbbells or spheres with radiating spokes	Common in horses	Neutral-alkaline
Calcium oxalate	Octagonal or cross-shaped	May be in small numbers common with antifreeze (ethylene glycol) toxicity	Acidic-neutral
Cholesterol	Parallelogram, corners indented, notched	Cell membrane lysis	Acidic-neutral
Cystine	Hexagonal plates	Cystinuria	Acidic-neutral
Drugs			
Ampicillin	Long and needlelike		
Sulfonamides	Round, dark spokes radiating outward		
Triple phosphate	Prism or elongated lid shape	No disease significance	Neutral-alkaline
Tyrosine	Colorless–pale yellow needles arranged in parallel	Liver disease	Acidic-neutral

Table 12-2. Urine sediment crystals.

PATIENTS

Cow No. 1—Holstein Cow

The Holstein cow had calved normally 10 days earlier, but had developed respiratory distress. The local veterinarian had administered penicillin and a triple sulfonamide preparation, but the cow did not seem to improve. The heart rate was 80 beats per minute and the body temperature was 40°C. The owner brought the animal to the nearest veterinary teaching hospital. X-rays revealed that the cow had severe pneumonia. Bloodwork showed that there was an elevated white blood cell count (response to an infection). Your colleague decided to treat the infection aggressively with amoxicillin trihydrate and gentamicin sulfate as intramuscular injections. The cow still did not improve, and she refused to eat. Dexamethasone and flunixin meguline were given for two consecutive days. The cow was tube-fed with fresh rumen liquor and dehydrated alfalfa. Replacement electrolytes were added to the cow's water and her condition improved until the 8th day following her admission. The cow is dehydrated and anorectic. Analysis of a blood sample reveals elevated serum creatinine levels and low serum calcium levels. Your team of specialists has been asked to consult with the attending veterinarian.

Hypothesis

Write a hypothesis that accounts for the dehydration and elevated serum creatinine.

How might the cow's condition be related to her course of treatment for pneumonia?

Obtain a urine sample for this animal from your teacher and perform all pertinent tests. Record your results in the table provided.

Horse No. 1—Palomino Gelding

A 6-year-old palomino gelding was turned out to pasture in early November. After several weeks, the horse developed **azotemia**, was weak and ataxic. The animal's appetite decreased and the owner brought the animal into the veterinary teaching hospital for treatment. You ask the horse's owner if the pasture has any trees in it. The owner replies that the pasture has two maple trees by the pond and several oak trees next to the fencerow. You inquire about herbicide use on the pasture, and the owner informs you that the pasture was sprayed for thistle and wild onion (see Table 12-3) 4 weeks before the horse was turned out on the pasture.

Drugs	Fungal toxins	Heavy metals
Antibiotics	Aflatoxin B_1	Arsenic
Aminoglycosides		Cadmium
Amphotericin B		Mercury
Cephaloridine		Uranium
Sulfonamides		
Polymyxin B		
Vancomycin		
Phenylbutazone		
Vitamins D_2, D_3, and synthetic K_3		

Internal substances	Toxin plants	Other
Hemoglobin (from blood) (cantharidin)	Black greasewood	Blister beetle
Myoglobin (from muscle)	Cultivated onion	Dioxin
	Day-blooming jessamine, day cestrum, and wild jasmine	Oxalates
	Locoweed and poisonvetch	
	Oak (including acorns)	
	Rayless goldenrod	
	Red maple	
	White snakeroot	
	Wild onion	

Table 12-3. Some substances that are toxic to the equine kidney (after Schmidt, 1988).

Why is it important to collect information about the pasture's vegetation?

How might chemicals such as fertilizers or herbicides play a role in the horse's disease process?

Hypothesis

Write a hypothesis that suggests a cause for the horse's renal failure and conduct a urinalysis on the sample provided by your instructor. Record the results in the table provided.

Cow No. 2—Jersey Cow

A 3-year-old Jersey cow calved 3 days earlier, and while the calf appears to be healthy, the cow's condition has deteriorated. The cow calved in the pasture unattended early in the morning and brought the calf to the barn later the same day. The cow has developed mastitis, severe dehydration, rumen stasis, and a high fever. Urine output has decreased. The cow has been admitted to the veterinary teaching hospital for evaluation and treatment.

What is the relationship between calving and mastitis?

Could the cow's fever be solely attributable to the mastitis?

Are there other possible explanations?

You ask the cow's owner if a veterinarian examined the cow and calf after the birth, and you are informed that no veterinarian was called because everything appeared to be normal.

Obtain a urine sample for Cow No. 2 from your instructor and perform any tests you think are necessary to ascertain the cow's problem. Record the results in the table provided.

Cow No. 3—Angus Bull

A 2-year-old Angus bull presents in the clinic teaching hospital with hemorrhagic diarrhea (bloody), rapid breathing, and ataxia. Serum glucose is elevated (121 mg/dl) and serum creatinine is high (5.0 mg/dl). You ask the owner if there have been any changes in the bull's feed (schedule, amount, type) or environment. The owner tells you that the only recent change is a new fence. You then inquire as to the type of fencing materials and the owner answers that the fence is constructed of pressure-treated lumber and recycled railroad ties (split and used as the posts). The owner then tells you that he has been unable to paint the fence because it has rained at least three times a week for the past 3 weeks since the fence was completed.

Why is it important to collect information on feed and animal housing when assessing the animal's condition?

Hypothesis

Suggest a relationship between the types of fencing material used, the amount of rain, and the bull's renal problems.

Urinalysis	General characteristics					Dipstick tests				Chemical tests		Urinary sediment		
Patient	Color	Odor	Clarity	pH	Specific gravity	Bilirubin	Heme RBCs	Glucose	Protein	Acid precipitation test-protein	Benedict's test-glucose	Casts	Crystals	Cells (type)
Cow no. 1 Holstein cow														
Horse no. 4 palomino gelding														
Cow no. 2 Jersey cow														
Cow no. 3 Angus bull														
Horse no. 2 Thorough-bred mare														

Obtain a urine sample for Cow No. 3 from your instructor and perform the appropriate urine analysis. Record the results in the table provided.

Horse No. 2—Thoroughbred Mare

A 5-year-old Thoroughbred mare is brought to the teaching hospital for treatment. The mare is in respiratory distress, is severely dehydrated, and appears to have some neurological impairment. You inquire about the feed and housing of the animal. The owner tells you that there have been no recent changes in the horse's stall, pasture, or fencing. The owner does tell you that she recently purchased a large amount of corn at a reduced price and that the symptoms appeared after the mare's feed was changed to include some of the corn.

Summary of Results

Fill in the table with the results of the tests that you have performed for each animal(s) assigned to your team.
How might the change in diet be related to the mare's condition?

Why would the mare be exhibiting neurological difficulties as a result of a dietary change?

Hypothesis

Write a hypothesis that relates the mare's condition to her feed. Obtain a urine sample for Horse No. 2 from your instructor and record the results in the table provided. Suggest a course of treatment and/or prevention for the horse's ailment.

With respect to animal husbandry practices, why is it important to examine the housing of the animals for any changes that might have caused the symptoms?

How might animal density in the pasture, barn, or holding pen affect the course of the disease process?

What considerations should a producer make when setting up a management schedule for his or her pasture?

What are some of the long-term animal health implications of "routine" medication doses for the maintenance of healthy livestock herds in high-density settings?

GLOSSARY OF TERMS

acetonuria: Presence of ketones (ketone bodies) in the urine resulting from systemic ketosis (production of ketone bodies).

albuminuria: Presence of albumin in the urine; symptomatic of kidney dysfunction.

anabolic: (*ana*—up) Energy-requiring reactions in which smaller molecules are chemically combined to form larger molecules.

azotemia: (*azo*—excessive) Higher than normal amounts of nonprotein nitrogen-containing molecules in the blood plasma resulting from metabolic processes; may also be called azoturia. May result from excessive breakdown of protein as seen in racehorses and draft animals following a hard workout; azotemia also occurs when about 75% of the renal tissue is damaged.

bilirubinuria: Presence of bilirubin in urine; may indicate kidney and/or liver dysfunction.

collecting duct: Portion of the nephron that follows the distal convoluted tubule and connects the nephron with the renal pelvis of the kidney; water reabsorbed from this region of the nephron is regulated by ADH (antidiuretic hormone).

cortex: Outer region of the kidney; primary function of the cortical region is the filtration of blood; glomerulus located in cortical region only.

distal convoluted tubule: (*distal*—far away from) Highly coiled tubule following the renal loop of the nephron which connects to the collecting duct, reabsorption of several electrolytes into the peritubular capillaries occurs in this region of the nephron.

electrolyte: A substance which separates when dissolved in water and which is capable of conducting an electric current; in blood plasma, important electrolytes are sodium (Na^+), potassium (K^+), chloride (Cl^-), calcium (Ca^{+2}), bicarbonate (HCO_3^-), and phosphate (PO_4^{-3}), sometimes also called *ions*).

glomerular filtration: Occurs in the cortical region of the kidney; glomerulus produces filtrate as liquid portion of the blood passes through filtration slits in the glomerulus. Glomerular filtrate contains: (a) plasma proteins, (b) electrolytes: Na^+, Cl^-, K^+, Ca^{+2}, and (c) nitrogen wastes: urea (toxic ammonia is converted to urea by liver).

glomerulus: (*glomer*—ball of yarn) Rounded ball of capillaries which is surrounded by a single layer of squamous epithelial cells (termed "Bowman's Capsule") in the nephron; functions in blood filtration.

glucosuria: Presence of glucose in the urine; may result from kidney, liver, and/or pancreatic dysfunction or abnormal eating patterns.

hematuria: Presence of whole red blood cells (erythrocytes) in the urine resulting from injury or disease.

hemoglobinuria: Presence of heme pigment in urine resulting from the breakdown of red blood cells due to inflammation, injury, and/or infection.

homeostasis: (*homeo*—same; *stas*—standing) Maintenance of stable internal environmental conditions, usually within a "normal" range; for instance, the normal range for human urine pH is 4.0–8.0 with an average value of 6.0.

hyperglycemia: (*hyper*—excess; *glyc*—glucose) Abnormally high levels of glucose in the blood plasma (serum).

loop of Henle (loop of the nephron): Consists of a *descending limb, loop,* and *ascending limb;* situated between the proximal and distal convoluted tubules of the nephron; water loss from the descending limb and sodium and chloride loss from the ascending limb result in urine concentration and water conservation.

medulla: Inner region of the kidney; the function of the medullary region is the collection of wastes produced in the cortical region; salt gradient in medulla results in the concentration of wastes in the filtrate; water and electrolytes reabsorbed from different regions of the nephron in the medulla.

myoglobin: (*myos*—muscle) An iron-containing red pigment molecule found in muscle.

myoglobinuria: (*myos*—muscle; *uria*—relating to urine) The presence of myoglobin in the urine; may be due to anorexia or excessive exercise.

pH: A measure of the relative concentration of hydronium (H^+) ions. The greater the H^+ concentration, the more acidic a substance is; the lower the concentration of H^+ ions, the more alkaline a substance is.

1		7		14
very acidic	acidic	neutral	alkaline	very alkaline

proximal convoluted tubule: (*proximal*—close to) Highly coiled portion of the nephron following the glomerulus and preceding the loop of the nephron. Water, electrolytes, and glucose are reabsorbed from this region of the nephron.

solute: Any substance dissolved in a liquid.

tubular reabsorption: Occurs in the limbs and loop of Henle, distal convoluted tubules; *counter current multiplier mechanism*:

1. Water is lost from the filtrate in the descending limb of Henle.
2. Salts (Na^+, Cl^-) is lost from the ascending limb.
3. Water is lost from the collecting duct and the distal convoluted tubule.

tubular secretion: Substances differentially removed from blood in peritubular capillaries (*peri*—around; *tubular*—convoluted tubules) and added to the glomerular filtrate (compounds such as penicillin, hormones, vitamins, etc.).

EXERCISE 13

Animal Reproduction

OBJECTIVES

Upon completion of this lab exercise, you should be able to:

- describe the process of gametogenesis including the similarities and differences between spermatogenesis and oogenesis.
- describe the hormonal mechanisms which regulate sperm production in a mammal.
- describe the hormonal regulation of the estrus cycle in a mammal.
- describe the processes of fertilization and cleavage.
- differentiate between:
  - blastula and gastrula.
  - blastopore and blastocoel.
  - blastocoel and archenteron.
- name the three embryonic germ layers and describe the tissues/structures which develop from each.
- distinguish between the four extraembryonic membranes in a vertebrate embryo and give the function of each in a bird egg.

Suggested Reading:

Chapter 11 of *The Science of Agriculture: A Biological Approach*, 4th edition

INTRODUCTION

Successful reproduction is essential to the maintenance of life on our planet. All organisms have some reproductive mechanism. One-celled organisms can reproduce by simply dividing to form two new cells. Reproduction by miotic cell division produces two organisms that are genetically identical to one another and to the parent cell from which they are derived. This type of reproduction is an example of *asexual* reproduction. Asexual reproduction is common in plants, fungi, protists, and bacteria. Some simple animals such as sponges and jellyfish are also capable of asexual reproduction. Higher animals such as vertebrates (animals that have a bony spine) cannot reproduce asexually. Animal reproduction involves the union of two cells from different individuals resulting in an offspring which possesses a unique genetic combination of maternal and paternal characteristics, *sexual* reproduction.

Meiosis, a special type of cell division, produces sex cells that unite to form the new individual. Meiotic cell division results in four new **haploid** cells, each with half the number of chromosomes found in the parent cell. This reduction in chromosome number is necessary; the new offspring resulting from egg and sperm fusion should have the same total number of chromosomes as either parent.

GAMETOGENESIS

Meiosis in Animals: Bull Sperm

The products of meiosis in animals, sperm and eggs, do not look alike. Egg cells are much larger than sperm cells. In this exercise you will observe several stages in the production of sperm. Meiosis produces haploid cells, and various intermediate stages of molecular reorganization and differentiation gradually elongate the round haploid cells to form mature sperm cells.

Obtain a prepared slide of the testes. Locate the seminiferous tubules, spermatogonia, primary spermatocytes, spermatids, and spermatozoa (sperm). The testis is the location of meiosis in the male. Spermatogonia are **diploid** cells at the outer edge of the seminiferous tubules, which divide by **mitosis** to form primary spermatocytes. Haploid sperm are produced from diploid primary spermatocytes when they undergo meiosis. A primary spermatocyte undergoes the first meiotic division to produce two secondary spermatocytes. Each secondary spermatocyte undergoes the second meiotic division so that the end result is four spermatids. The spermatids then differentiate, developing the specialized structures of spermatozoa or mature sperm (see Figure 13-1). Using the high-power objective lens, identify the spermatogonia, primary spermatocytes, and spermatids on the slide provided by your instructor. Observe the structure of sperm on the microscope using the oil immersion lens (1,000×).

HORMONAL REGULATION OF SPERMATOGENESIS

There are three hormones that are directly involved in the process of spermatogenesis: **Follicle-stimulating hormone (FSH)**, **luteinizing hormone (LH)**, and **testosterone**. FSH and LH are both produced by the anterior (front) of the pituitary gland, which is located at the base of the brain. FSH and LH both act on the seminiferous tubules of the testes to stimulate spermatogenesis. LH is also active on the specialized cells inside of the testes, which are responsible for secreting the male hormone testosterone. Testosterone, produced by the testes,

Directions to the student: Label Figure13-1 using the following terms—**spermatogonia, primary spermatocytes, secondary spermatocytes, spermatids** and **spermatozoa** (sperm). Color the haploid cells blue and the diploid cells red.

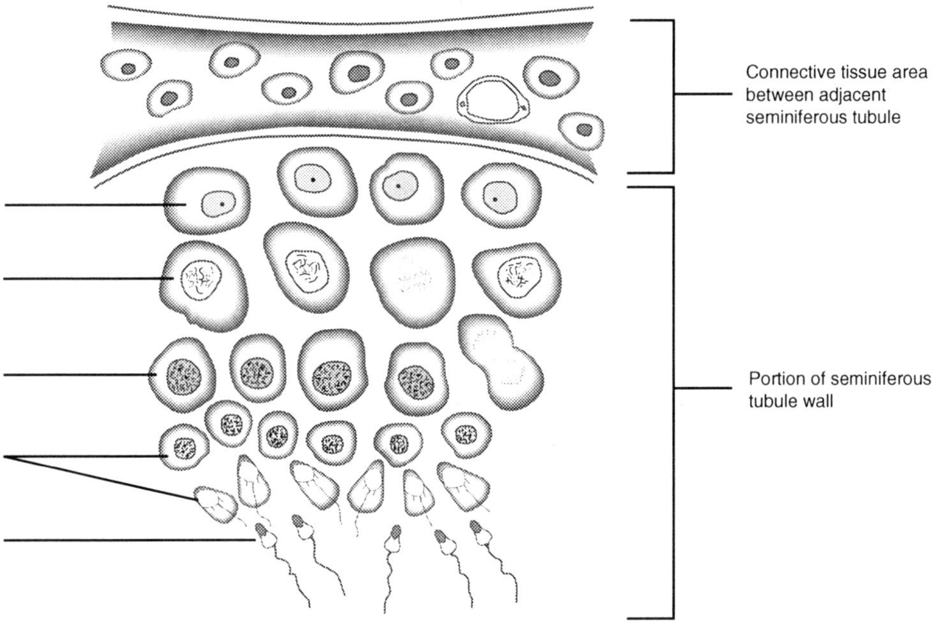

Figure 13-1. Sperm formation in male testes.

is responsible for the secondary sexual characteristics associated with "maleness." Castrated male calves do not develop the same muscling patterns as young bulls even when their diets are identical. Testosterone increases muscling by affecting the animal's protein metabolism.

OOGENESIS

The meiotic divisions that produce the **ovum** (mature egg) are different from those which produce sperm—instead of four haploid daughter cells, only one large viable cell is produced, the egg. The divisions of the cell's cytoplasm accompanying oogenesis are unequal, with the majority of the cytoplasm surrounding the egg's nucleus. The other three cells, polar bodies, degenerate shortly after forming. A diploid oogonium undergoes mitosis to form a primary oocyte (2N). The primary oocyte undergoes the first meiotic division forming a polar body (N) and a secondary oocyte (N). The haploid secondary oocyte completes the second meiotic division to from an additional polar body (N) and a haploid ootid. The ootid will eventually develop into a mature ovum (egg) (see Figure 13-2). When a female mammal is born, all of the eggs she will ever shed during her **estrus cycles** are in her ovaries. As an animal ages, the advancing age of her eggs increases the risk for birth defects, particularly those

Directions to the student: Match the numbers in Figure 13-2 to the following terms or events. Write the letter which best corresponds to the term or event in the blank provided.

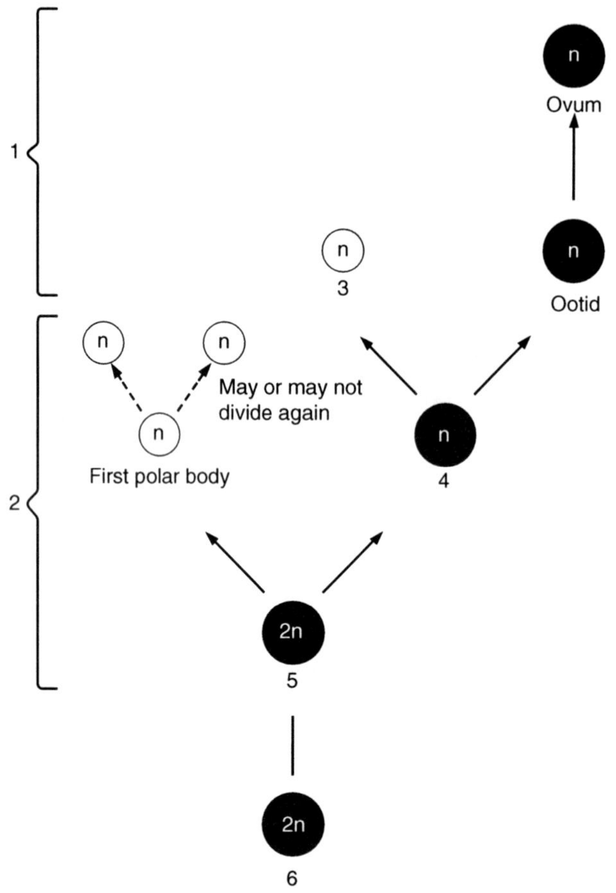

_____ Maturation

_____ Meiosis

_____ Oogonium

_____ Primary oocyte

_____ Secondary oocyte

_____ Second polar body

Figure 13-2. Oogenesis.

birth defects or conditions attributable to chromosomal abnormalities. The estrus cycle is the female reproductive cycle occurring at regular intervals throughout the female animal's life. Different domesticated animals have estrus cycles with different characteristics. Effective herd management cannot be accomplished without a good understanding of the reproductive traits of the breed (see Table 13–1).

Examine the slide of the ovary provided by your instructor using the 10× objective. Some of the slides contain primordial follicles, very small follicles containing immature eggs near the edge of the ovary. Slightly larger immature follicles are termed primary follicles. Secondary follicles are some larger fluid-filled structures with a visible immature egg. The

	Cow	Mare	Ewe	Doe	Sow
Age at puberty (in months)	8–12 (breed differences)	15–24 (seasonal effects)	7–10	6–8	5–8
Age at sexual maturity (in months)	30	36	10	8 (if born early in year)	10
Breeding season	all year	April–September; all year with lights	Breed variation from fall to all year	September–January in Northern Hemisphere	All year
Estrus cycle length	21 days	21 days	17 days (14–19, range)	21 days	21 days
Estrus duration	12–18 hours	4–7 days	36 hours (24–48, range)	18–36 hours	48–72 hours
Optimal breeding time (after estrus onset)	10–16 hours	48–72 hours	18–24 hours	24–36 hours	12–30 hours
Time of ovulation (after onset of estrus)	4–16 hours	24–48 hours before estrus end	24 hours	12–36 hours	24–42 hours
Ovulation rate (number of eggs)	1	1	1–2	2–3	10–20
Implantation (days)	10–12	25–56	14–18	10–11	11–16
Gestation length (days)	278–293	330–345	144–151	146–151	112–115
Diploid chromosome number	60	64	54	60	38
Time to rebreed	45–90 days after calving	25–30 days after foaling	First estrus	First estrus	First estrus

Table 13-1. Reproductive parameters of female domesticated animals

cells surrounding the secondary follicle are responsible for the secretion of **estrogen**, the female sex hormone. Identify these structures on the slide of an immature ovary and label them in Figure 13-3. Obtain a Graafian follicle slide from your instructor and use the 10× objective

Directions to the student: Label the structures in Figure 13-3 using the following terms—**corpus luteum**, **follicular fluid**, **ovulation**, **primary follicle**, and **secondary follicle**.

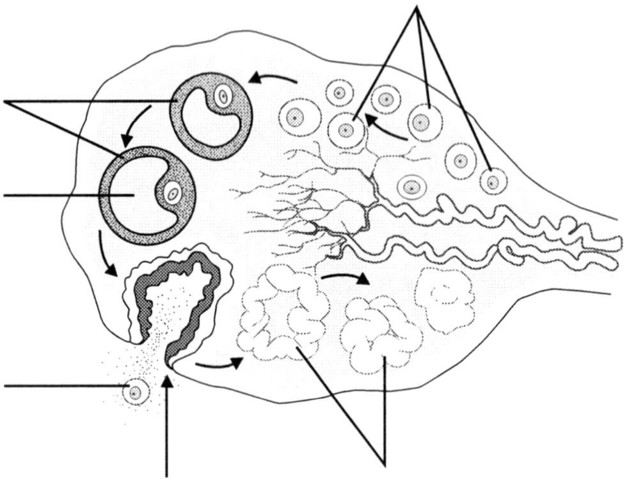

Figure 13-3. The structure of the ovary.

to observe the mature (Graafian) follicle. The egg inside is ready to be shed during ovulation. If the egg is fertilized, the cells surrounding the Graafian follicle continue to produce hormones that prevent the shedding of the uterine lining. These secretory cells are called the corpus luteum. The corpus luteum will continue to secrete hormones maintaining pregnancy until the placenta begins to make hormones at high enough levels to prevent spontaneous abortion. Label the ovum and the corpus luteum in Figure 13-3.

HORMONAL REGULATION OF THE ESTRUS CYCLE

The hormonal regulation of the female reproductive (estrus) cycle is far more complex than in the male animal. The anterior pituitary hormones LH and FSH are secreted in large quantities prior to ovulation. Blood levels of estrogen are also at peak levels prior to the release of the egg from the ovary. When blood levels of estrogen, LH, and FSH drop at the same time, ovulation occurs (see Figure 13-4 for an example). Following ovulation, **progesterone** is secreted in large amounts to prepare the lining of the uterus to receive the fertilized egg. The highest progesterone blood levels occur around the time of implantation, preventing the loss of the developing zygote.

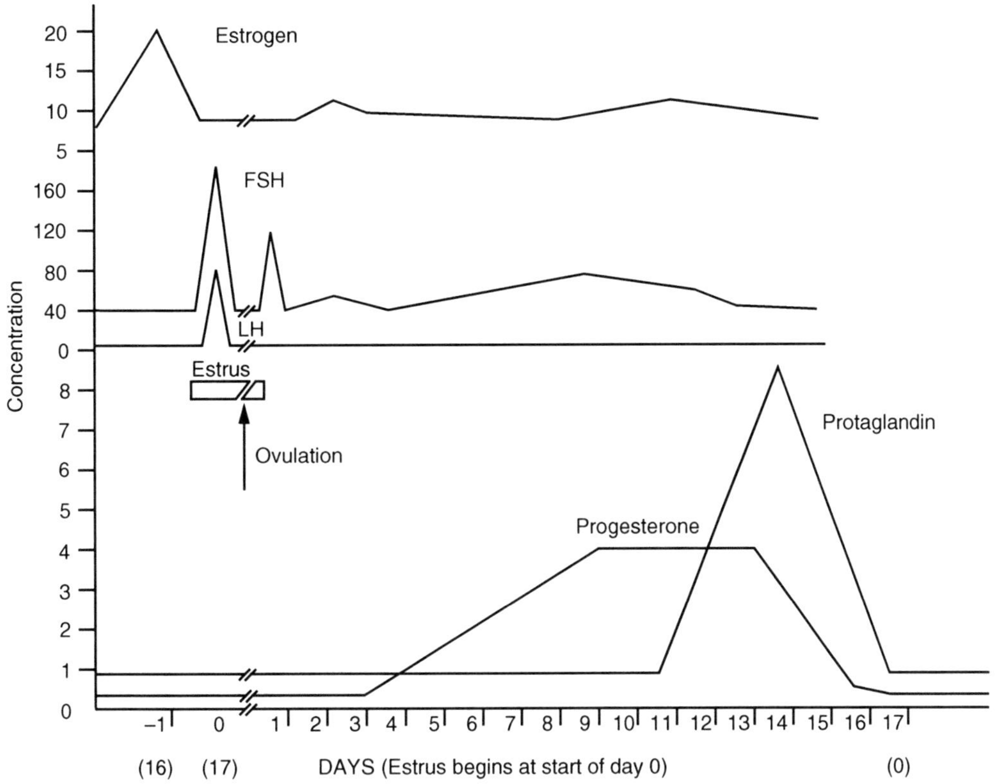

Figure 13-4. Blood hormone levels during the sheep estrus cycle.

GENERAL EMBRYOLOGICAL DEVELOPMENT

Embryology is the science of studying embryos. Many invertebrates (animals without backbones) and all vertebrates begin their existence as a fertilized ovum, or zygote. Embryological development includes the processes of gamete formation, gamete union to form the zygote, their subsequent development from a zygote to all the complicated tissues and organs of the body and their integration into an individual.

You will follow embryonic development from fertilization until development establishes the basic body plan of the two different animals in this exercise. The following paragraphs describe critical developmental stages and processes. However, you should note that development is a continuous process.

Cleavage is the miotic division of the fertilized ovum into 2, 4, 8, 16, 32 cells and so on through the production of a hollow sphere of cells, the **blastula** (*blast*–to germinate, to segment). The cells comprising the walls of the blastula are the blastomeres (*mere*–a part). This wall is the first germ layer, the primitive **ectoderm** (*ecto*–outer, *derm*–skin or layer). The cavity of the blastula is the **blastocoel** (pronounced blast-oh-seal).

The blastula stage is relatively short in duration. Cell division continues uninterrupted and rapidly to the second stage, the **gastrula** (*gaster*–stomach or gut). Gastrulation results in the formation of the second germ layer, internal to the first, with an opening to the outside. This second layer is the primitive **endoderm** (*endo*–irmer; *derm*–layer). The primitive endoderm is derived directly from the primitive ectoderm by *imagination,* in which the cell layer pushes in at one end, much like a deflated basketball. The process is aided in many animals by overgrowth of cellular layers, by cellular migration, and by cellular rearrangement, including thickening and thinning of layers. The cavity inside the tube of primitive endoderm is the **archenteron** (primitive gut). With the formation of the first two layers, ectoderm and endoderm, the essential precursors of a multicellular animal are established.

In higher animals, including echinoderms (e.g., sea urchins, starfish) as well as the vertebrates, a third stage follows swiftly upon the second. A third germ layer, **mesoderm** (*meso*–middle), forms between the primitive ectoderm and endoderm. This process lacks a specific name.

The ectoderm gives rise to the nervous system and the skin. The endoderm develops into the digestive and respiratory systems and structures such as the liver and pancreas. The mesoderm develops into the skeletal, muscular, reproductive, circulatory, and excretory systems. This process is identical to a point in all vertebrates and echinoderms (see Figure 13-5).

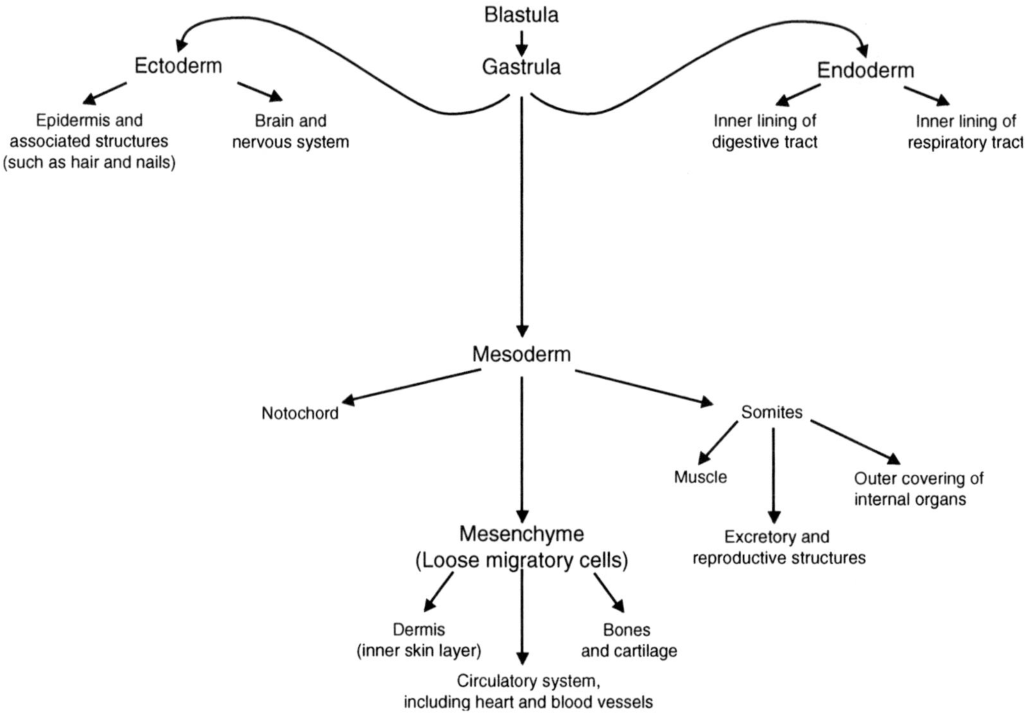

Figure 13-5. The derivation of body tissues from the three embryonic layers.

SEA URCHIN DEVELOPMENT

The early embryonic development of sea urchins, starfish, and other echinoderms is identical to the sequence of events in higher animals. Since echinoderms are inexpensive and easy to obtain, students of developmental biology frequently use them.

The Egg and Fertilization

(See Figure 13-6 for the letters referenced for each event.)

1. Fertilization is external in sea urchins.
2. First the male gamete (sperm) fertilizes the egg, forming the zygote (B). Note that the zygote possesses animal and vegetal hemispheres, which form an animal-vegetal (AV) axis.
3. The first cleavage in the echinoderm egg occurs in a vertical plane, parallel to the animal-vegetal (AV) axis. What is the relative size of the first two blastomeres formed by this cleavage?
4. The second cleavage is perpendicular to the first cleavage furrow but still parallel to the AV axis (D).
5. The third cleavage is at right angles to the first two (E).
6. Successive cleavages result in a ball of small cells called a morula (G). Note that all the cells are still the same size. Does the total mass of the living state increase during this early cleavage?

Blastula

(See Figure 13-6 for the letters referenced for each event.)

The solid ball of cells of the morula stage hollows out so that the organism now consists of a central cavity the blastocoel, surrounded by a single layer of cells. This stage is called the blastula (I). Has there been a change in cell size? Cell number? Sea urchin embryos at the blastula stage have cilia and are therefore motile (J).

Gastrulation

(Refer to Figure 13-6 for the letters referenced for each event.)

1. Gastrulation begins as the cells of the vegetal pole push inward against the inside wall of cells on the opposite side, forming a cup-shaped embryo, the gastrula. The cavity resulting is the primitive gut (archenteron). The opening to this cavity is called the **blasto-pore** and is destined to become the anus. (L)

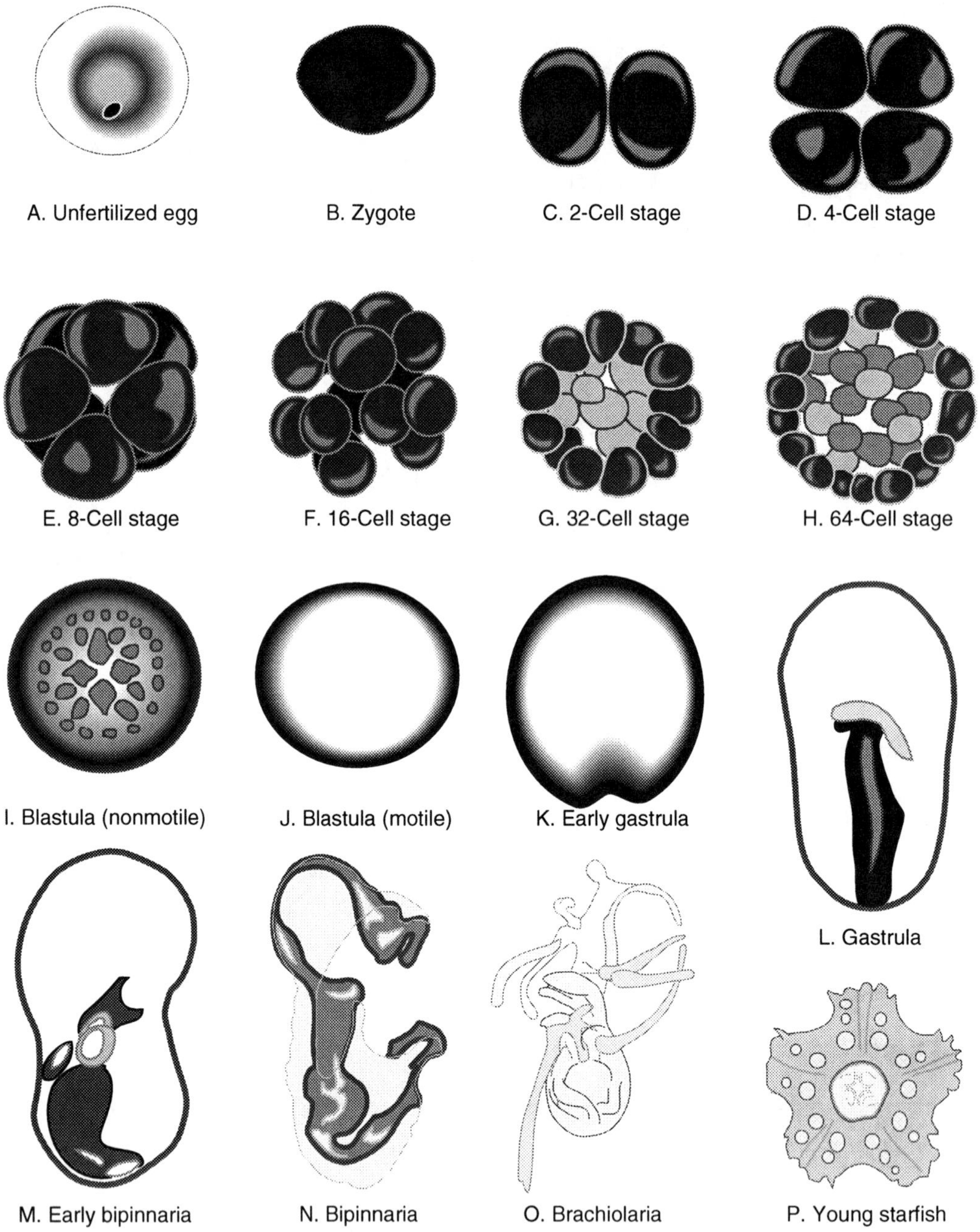

Figure 13-6. Starfish embryological development—similar in the sea urchin.

2. The early gastrula has two layers of cells, one lining the archenteron and the other forming the outer wall of the embryo. In the space between these two layers, a third layer of cells develops, resulting in a three-layered embryo. (K)
3. Cells continue to divide and to differentiate, forming a bipinnaria larva. (N)
4. Two to three weeks after fertilization, the larva undergoes **metamorphosis** into a young adult sea urchin.

PROCEDURES

In this lab activity, you will observe a sampling of the events leading to the adult stage. You will not learn all of the details; nevertheless, you should obtain some insight into the principles of development. In this laboratory exercise you should look at the following:

1. Fertilization and cleavage in the sea urchin.
2. Prepared microscope slides of preserved and stained starfish and frog embryos.
3. Prepared slides of a 2-day-old chick embryo, plastimounts and live examples of 4-day-old embryos, and live examples of 6-, 8-, and 10-day-old embryos.

SEA URCHIN FERTILIZATION

Sea urchin sperm and eggs are harvested by injecting KCl solution into the sea urchin body cavity causing gamete release. Each pair of students can make a slide to observe fertilization and the first cleavage. To observe sea urchin fertilization, follow the instructions below.

1. Have the instructor pipe some egg suspension into the center of a depression slide.
2. Look at the slide.
3. Pipette sperm onto a depression slide.
4. Place a cover slip over the depression; do not compress. Observe it under low power (10×objective) and high power (40×objective). Note the jelly and egg cortex of the unfertilized egg. You may be able to see cortical granules if you slightly close the iris diaphragm of the microscope. You should be able to see many unsuccessful sperm attached to the exterior of the egg.
5. Following fertilization, the cortical granules disappear. The vitelline membrane becomes visible around the egg. This membrane will form the fertilization membrane.
6. After viewing fertilization and the formation of the fertilization membrane, *turn off the light* of the microscope. Make sure it does not dry out (renew the fluid level with salt water) and view it later in the lab. You should see developmental stages according the following schedule:

Fertilization	0 min.
Fertilization membrane	2 to 5 min.
1st cleavage (2-cell)	50 to 70 min.

(*Continued*)

2nd cleavage (4-cell)	78 to 107 min.
3rd cleavage (8-cell)	2 hours
Blastula	6 to 8 hours

7. Make drawings and notes on these stages. Observe the early cleavage stages of eggs fertilized several hours before the class period.
8. Wash and dry depression slides at the end of the lab and return them to the box of slides.

ECHINODERM DEVELOPMENT

1. Obtain prepared slides of starfish development (if sea urchin fertilization does not occur).
2. Note that the early developmental stages of the echinoderm (starfish and sea urchin) are reasonably transparent because they have comparatively little yolk. The slides available for your use contain either whole embryos or sections of embryos at all stages of development through the larval stage. Study the embryos with the low power (10×) of the compound microscope.
3. Find all the developmental stages from fertilized eggs to young larvae (see Figure 13-6).
4. Draw these stages in the "Results and Discussion" section.

CHICKEN DEVELOPMENT

Follow the stages of chicken development in Figures 13-7 through 13–9, and Figure 13-11.

The Egg and Fertilization

1. The cell, or ovum, released from the hen's ovary consists of the yellow yolk mass. After copulation, sperm swim to the upper end of the oviduct and unite with the ovum. The egg nucleus undergoes its second meiotic division after a sperm has penetrated the blastodisc.
2. Yolk is so abundant in the bird egg that cleavage is restricted to a small disc of cytoplasm at the animal pole of the egg cell. After fertilization, cell division partitions this yolk-free cytoplasm to form a cap of cells called the blastodisc which rests on the large, yolky, undivided portion of the original egg cell. Cleavage is followed by a separation of the blastodisc into an upper and lower layer. The cavity between these two layers is the blastocoel. This embryonic stage

is the equivalent of the blastula, although its form is different from the hollow ball of frog or sea urchin blastulas (see Figure 13-7).

3. The upper layer of the blastoderm becomes the ectoderm and the lower layer becomes the endoderm. Following the formation of the blastoderm, an involution of the cells along the midline of the embryo produces a narrow grove, the primitive streak. Cells of the ectodermal layer then migrate toward the primitive streak, turn under, and spread out as a mesodermal layer between the ectoderm and endoderm. When the mesoderm is fully established, gastrulation is complete. Eventually, the surface ectoderm will be folded under the embryo. When this folding is completed, the ectoderm will no longer be a single sheet of cells but a complete outer covering.
4. In addition to the embryo itself, the primary germ layers also give rise to four extraembryonic membranes that support further embryonic development (see Figure 13-8).
   a. The **yolk sac** grows over and around the yolk and gets smaller, gradually, as yolk is used up during the growth of the embryo.
   b. The **amnion** surrounds the developing embryo everywhere except on the ventral side. It holds a fluid that bathes the embryo.
   c. The **allantois** lies against the eggshell, just inside the chorion. This membrane is the breathing structure of the embryo. Gas exchange occurs between numerous blood vessels in the allantois and air outside the shell.
   d. The **chorion** lies just inside the shell and prevents excessive evaporation of water through the shell. It forms an outer enclosure around the embryo and the other three membranes.
5. During its passage through the oviduct, several accessory features are added to the ovum. In the first section of the oviduct, viscous

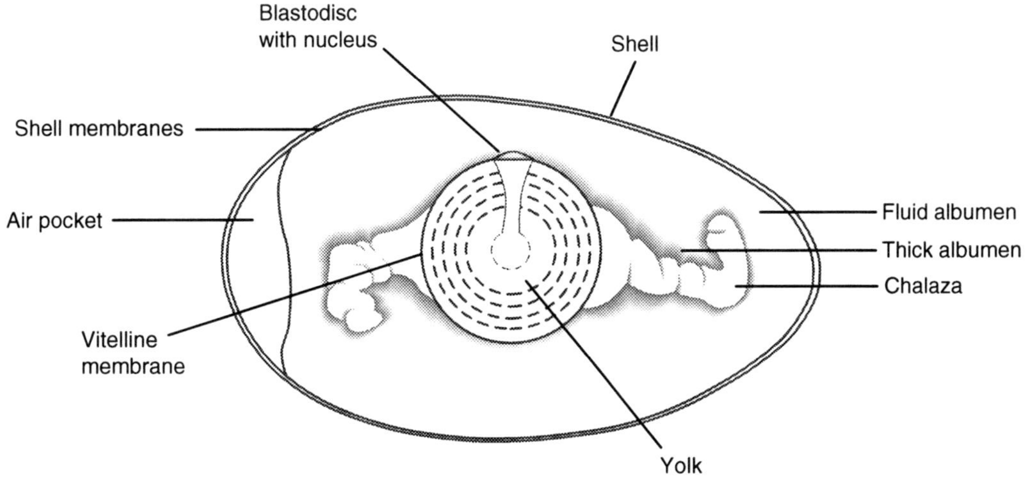

Figure 13-7. Internal anatomy of a bird's egg.

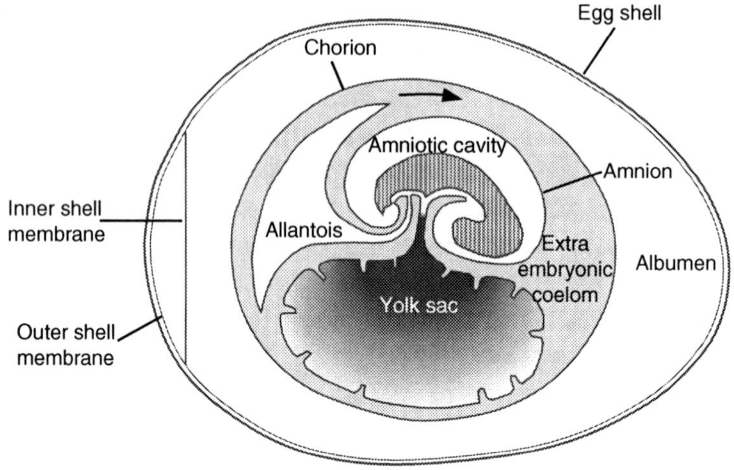

Figure 13-8. The arrangement of the four extraembryonic membranes in a chicken egg following embryo development.

stringy albumen is secreted and adheres to the ovum. In the next oviduct region, more watery albumen is added. This protein-rich solution (the egg white) provides nutrients for the growing embryo in addition to the yolk. Finally, two shell membranes form and a calcareous shell is applied to the outer albumen. By the time the egg is laid (about 20 hours after fertilization), the embryo is at the early gastrula stage.

VIEWING THE 2-DAY (48-HOUR) EMBRYO

Obtain a prepared slide of a 2-day-old embryo. You must use a dissecting microscope to view this. Note the following. (Figure 13-9 shows a 33-hour embryo.)

a. The embryo is twisted on its right side, and its head bends like a question mark (in reverse).
b. The optic vesicles, which become the eyes, are easy to see on either side of the head.
c. Posterior to the optic vesicles are the otic vesicles at each side of the hindbrain, which develops into the inner ear.
d. The amnion has begun to fold over the head and will envelop the whole embryo on another day of development.
e. The heart sends blood forward through three pairs of aortic arches, and then posteriorly through the dorsal aorta toward the tail to the yolk sac. Many vessels branch over this sac, and absorb nutrients and acquire oxygen from it before returning blood via the vitelline veins for another cycle through the heart and the embryo.

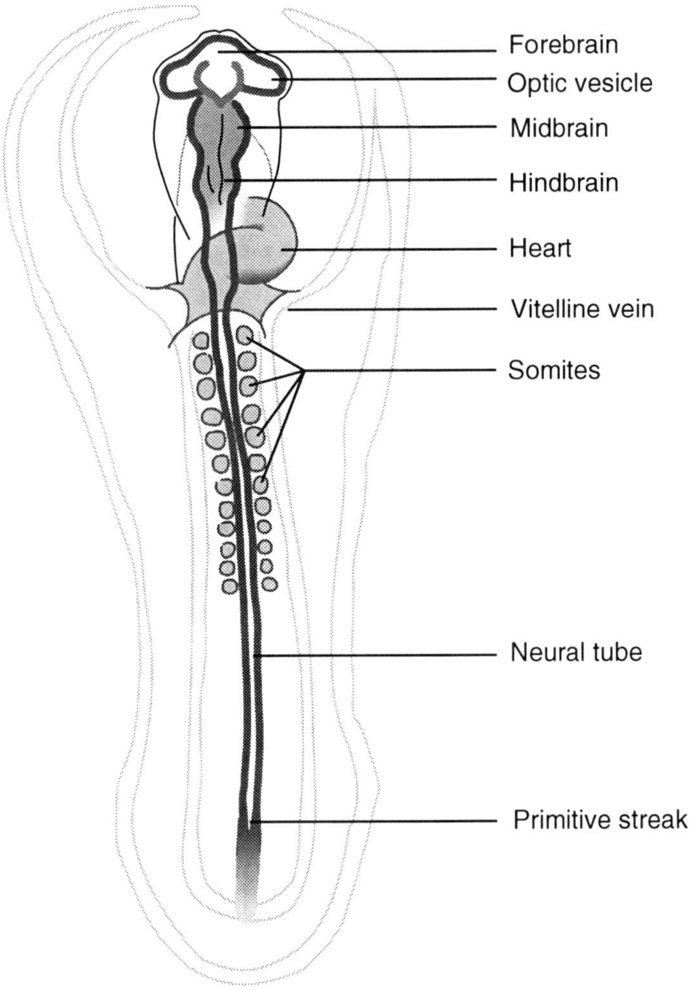

Figure 13-9. A dorsal view of a 33-hour chicken embryo.

f. The chick embryo should contain 18 pairs of **somites**. Can you count them on either side of the spinal cord? Begin at the hindbrain and count toward the tail. No limb buds are visible at this time.

Opening a 4- and 6-day-old Chick Egg

Eggs at this stage contain so much yolk that it is difficult to view the embryo. Therefore, you will remove the embryo from the surrounding membranes to observe it. Obtain an egg and follow the instructions below to open it (see Figure 13-10).

1. Make a "nest" for the egg with a paper towel in a finger bowl so that the egg rests securely in the bowl. Place the egg in the orientation shown. Allow it to remain there at least *2 minutes* before proceeding.

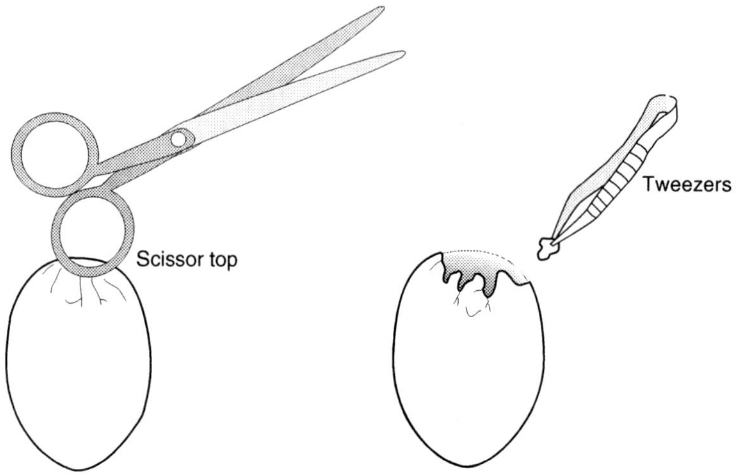

Figure 13-10. Procedure for opening a chicken egg.

2. Obtain scissors, forceps, spoon, and a watch glass containing warm physiological saline.
3. Crack the large end of the egg with the handle of the scissors.
4. Carefully and slowly clip the shell completely around the egg as indicated, and use the forceps to even the edges of the shell.
5. Remove the embryo from the yolk and place it in a dish of physiological saline for further examination. To do this, snip the area of the disk around the embryo and then move a spoon under it. Use scissors to separate the embryo from the albumen and the yolk. Then pour the embryo gently into the watch glass containing saline solution.

Viewing a 4-day Embryo

Locate in a live 96-hour embryo the structures labeled in Figure 13-11. Note that by the time the chick has undergone 96 hours of incubation, the following major events have occurred:

a. The embryo lies with its entire left side toward the yolk.
b. The wing and leg buds are now reasonably well-developed structures.
c. Between the hind limb buds, the allantois projects ventrally. It is one of the four extraembryonic membranes and functions in waste removal.
d. The chorion, formed at the same time as the amnion, serves with the allantois in the transport and exchange of respiratory gases. Note that the allantois expands beneath the chorion. Differentiation continues until (and after) the egg hatches.

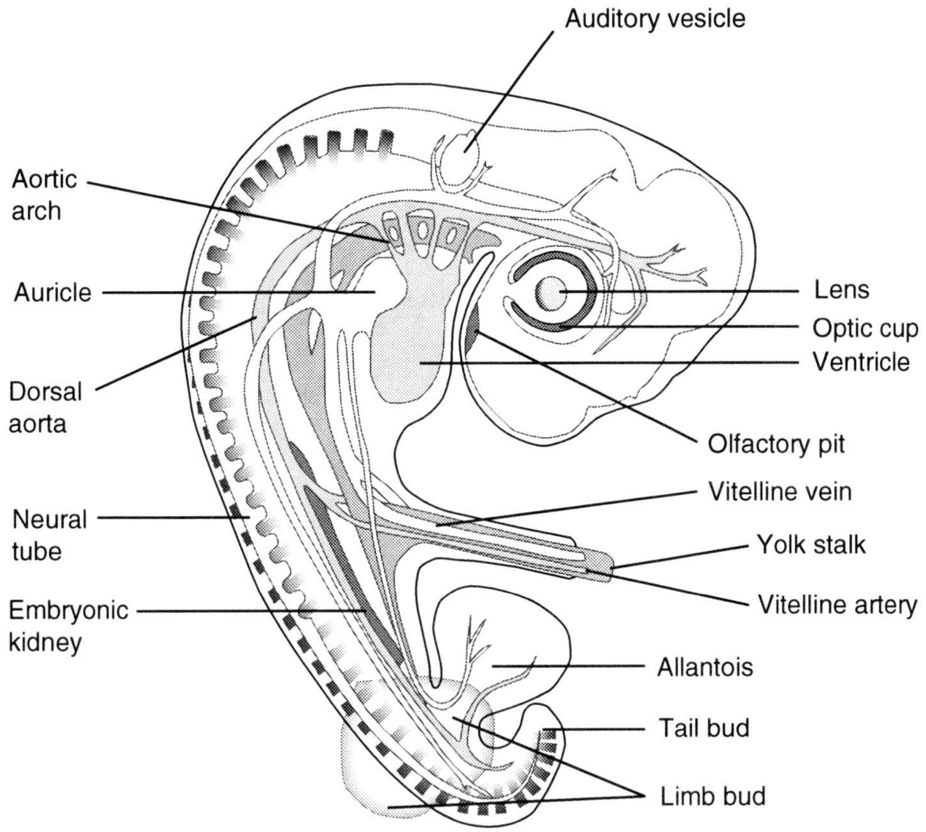

Figure 13-11. A side view of a 96-hour chicken embryo.

Viewing the 6-day Embryo

By the time the chick has undergone 6 days of incubation, the following major events have occurred:

a. The wing and limb buds have developed; grooves appear where the digits will develop.
b. The beak begins to develop.
c. The auditory vesicle becomes more prominent.

Opening an 8- and 10-day Chick Egg

Note that as eggs develop, the volume of yolk decreases relative to the volume of the embryo. (Why?) This allows us to remove the chick from the egg with its membranes intact.

Working in groups of four to five students, open an 8- or 10-day-old chick, following these directions.

1. Make a "nest" for the egg with paper towel in a finger bowl so that the egg rests securely in the bowl. Place the egg in the nest, in the orientation shown. Allow it to remain in that position for at least 2 minutes before proceeding.

2. Use a dissecting needle or probe make a small hole in the shell or crack the large end of the egg with scissors or scalpel handle.
3. Use forceps or fingers to pick away the shell. Initially avoid breaking the shell membrane and observe its position adjacent to the shell. Then carefully puncture the membrane.
4. Gently pour the entire embryo out into the finger bowl of warm physiological saline.
5. The chick can be seen within the chorio-allantois. Remove the membrane for further examination.

Viewing the 8-day Embryo

Obtain and observe the structure of an 8-day embryo, and note the following.

a. The eyes are large and conspicuous.
b. The beak and toes are distinct.
c. "Fingers" are appearing on the wing bud.

Viewing the 10-day embryo

By 10 days, the skin has entirely covered the embryo, and feather germs populate all but the extremities. Eyelids nearly cover the eyes and its black retinas. Opening just behind each eye are the ears. Wings and legs display conspicuous digits, and claw primordia protruding from the digits of each leg may be discernible. Internal organs, such as the heart, lungs, liver, kidney, and gut lie entirely in the embryo.

RESULTS AND DISCUSSION

1. Draw and label major structures of the sea urchin/starfish development at:

   blastula

late gastrula

neurula

2. Look at a prepared slide of a 2-day-old chick embryo under a dissecting microscope. Draw and label the following structures: optic vesicle, otic vesicle, heart, **neural tube**, and somites.

3. Draw in detail the 4-day-old chick embryo. Label eye, brain, heart, and allantois.

4. Use the provided space to make comparisons of the 6-day, 8-day, and 10-day embryos.

6-day

8-day

10-day

5. What adaptation does an avian (bird) egg have for survival in a terrestrial environment?

6. List the four extraembryonic membranes of a chicken egg and state the function of each.

7. Why would an understanding of the biology of the bovine estrus cycle be important to a cattle breeder?

8. What are some potential uses for synthetic hormones in the management of herds? (*Hint:* Think about the hormonal regulation of both male and female animals.)

GLOSSARY OF TERMS

allantois: (*attos*—sausage; *eidos*—form) An embryonic membrane in the amniote egg which serves as a respiratory surface and as a waste storage organ; in placental mammals, the allantois forms most of the umbilical cord, and hence plays an important role in placental development.

amnion: In amniote eggs and placental mammals, the amnion is the innermost extraembryonic membrane forming a protective fluid-filled sac that surrounds the developing embryo.

archenteron: (*arche*—beginning; *enteron*—gut) The internal cavity of the vertebrate gastrula lined with cells of endodermal origin; opening will form the opening of the digestive tract.

blastocoel: (*mastos*—svrout; *koilo*—hollow). The hollow, fluid-filled internal cavity of the vertebrate blastula.

blastopore: (*blastos*—sprout; *poms*—path or passage) During the early embryonic development of vertebrates, the point at which the blastula invaginates to form a gastrula; the blastopore, then, is the opening to the archenteron. In some animal groups, the blastopore becomes the mouth, in others (vertebrates, for example), the blastopore becomes the anus.

blastula: (*blastos*—sprout) A stage in vertebrate embryonic development consisting of a hollow ball of cells one cell layer in thickness.

chorion: The outermost extraembryonic membrane in amniote eggs and placental mammals. In the amniote egg, the chorion serves as a site for gas exchange and prevents the desiccation of the egg contents. In placental mammals, the chorion develops into the embryonic portion of the placenta.

cleavage: Rapid cell division that follows the fertilization of a vertebrate egg; continues until the blastula is formed.

diploid: (*diploos*—double; *ploion*—vessel) The condition of having two sets of chromosomes (2N); in animals, twice the number of chromosomes found in sex cells; in plants, the number of chromosomes characteristic of the sporophyte stage.

ectoderm: (*ecto*—outside; *derma*—skin) One of the three vertebrate germ layers that form during the gastrula stage of embryonic development; gives rise to the outside body covering (skin, hair, nails, etc.), the brain, the spinal column, and sensory organs.

endoderm: (*endon*—within; *derma*—skin) One of the three vertebrate germ layers that form during the gastrula stage of embryonic development; gives rise to the lining of hollow internal body organs such as the lining of the digestive, respiratory, urinary, and reproductive tracts, and forms some endocrine glands such as the pancreas.

estrogen: Female sex hormones that are produced by the ovaries; estrogens are responsible for the development of secondary female sexual characteristics and the development/regulation of female reproductive structures.

estrus cycle: The female reproductive cycle which results from hormonal action on the female reproductive organs.

follicle-stimulating hormone (FSH): Anterior pituitary hormone which stimulates the development of eggs and estrogen production in female mammals and initiates sperm production in males.

gamete: Sex cells that are produced in animals by meiosis, reductive cell division; gametes are haploid cells.

gastrula: (*gastro*—little stomach) In vertebrates, a stage of early embryonic development in which the one-cell layer blastula becomes a three cell-layer thick structure (*ectoderm, endoderm,* and *mesoderm*) with a hollow internal cavity which opens to the outside.

haploid: (*haploos*—single; *ploion*—vessel) Possessing one set of chromosomes (N or IN); formed following meiosis of diploid (2N) cells; in animals, sex cells are haploid; in plants, characteristic of the gametophyte stage; some protists and fungi are also typically haploid.

luteinizing hormone (LH): Anterior pituitary hormone which stimulates testosterone production in male mammals. In females, LH controls a variety of functions including the stimulation of ovulation, regulation of progesterone secretion by the corpus luteum, and preparing the milk glands (mammary glands) for milk production.

meiosis: (*meioun*—dimmish) A type of cell division in which the chromosome number is reduced by one-half, one of each pair of homologous chromosomes passes to each daughter cell; usually occurs in gamete-producing cells.

mesoderm: (*mesos*—middle; *derma*—skin) One of the three vertebrate germ layers that form during the gastrula stage of embryonic development; gives rise to the muscle, bone, and other connective tissue including the peritoneal cavity and the circulatory system.

metamorphosis: (*meta*—change; *morphe*—form; *osis*—state of) Pronounced change in body form following embryonic development; in amphibians, aquatic larval stage ⟶ tadpole undergoes metamorphosis to form semi-terrestrial adult ⟶ frog.

mitosis: (*mitos*—thread) A type of cell division in which each daughter cell is genetically identical to the parent cell; the chromosome number remains constant in miotic cell division.

neural tube: (*neuron*—nerve) In the vertebrate embryo, the neural tube is formed by the fusion of opposite longitudinal ectodermal folds.

notochord: (*noto*—back; *chorda*—cord) In chordates (members of the phylum *Chordata),* the notochord is a dorsal flexible cartilaginous rod which extends the length of the body and provides support as a primitive skeleton; the notochord is present at some stage of development in all chordates; in vertebrates, the notochord is present only in the developing embryo and is replaced by a bony vertebral column.

ovum: Female gamete.

progesterone: A female sex hormone produced by the ovaries which prepares the uterus to receive a fertilized egg and readies the milk glands for milk secretion.

somites: (*soma*—body) Somites are segments of mesodermal tissue which form during the differentiation of a vertebrate embryo.

testosterone: A male hormone produced by specialized cells in the male testis which controls growth and development of male reproductive organs, sperm, the body, and all secondary sexual characteristics.

triploblastic: (*tri*—three; *blastos*—sprout) Possessing three embryonic germ layers: ectoderm, endoderm, and mesoderm.

yolk plug: In the gastrula stage of vertebrates, the pigmented cells which fill the blastopore.

yolk sac: An extraembryonic membrane in the amniote egg and placental mammals; in the amniote egg, the yolk sac surrounds the egg yolk; in placental mammals, the yolk sac is empty and forms a portion of the umbilical cord.

EXERCISE 14

The Use of Natural Controls in Weed Management

OBJECTIVES

Upon completion of this lab exercise, you should be able to:

- describe the agricultural impact of weeds.
- identify growth characteristics which result in the classification of a plant as a weed.
- distinguish between:
  - monocot and herbaceous weed.
  - herbaceous and woody weed.
- test the effect of a readily available herbicide, 2,4-D or other.
- appropriate herbicide, on a monocot and a dicot herbaceous plant.
- relate agricultural practices to the spread/increase in weed density including:
  - tillage practices.
  - crop rotation practices.
  - harvest practices.
- describe the action of biological weed control measures including:
  - introduction of natural enemies for the particular weedy species.
  - use of plant derived and/or microbial substances for weed control.
  - use of cover crops in crop rotation.
- design and test a natural/biological control plan for a weed plant.

Suggested Reading:
Chapter 14 of *The Science of Agriculture: A Biological Approach*, 4th edition.

INTRODUCTION

Weeds have been problematic for growers for as long as agriculture has been practiced. It is estimated that 41% of the total dollar amount spent on controlling crop pests is spent on weed control annually (Hoagland, 1990). **Herbicides**, compounds that slow or arrest plant growth, constitute the largest proportion of global pesticide sales and the greatest pesticide usage category (pounds/acre) in the United States (Hoagland, 1990). Crop losses in the United States due to weedy plants

Crop	Average annual loss (× $1,000.00)
Field crops	6,408,183.00
Forage seed crops	37,400.00
Fruits and nuts	441,449.00
Vegetables	619,072.00
Total	$ 7,506,104.00

Table 14-1. Average annual losses in the United States between 1975–1979 for several different crop groups (estimated).

are staggering (see Table 14-1). Profitability in agriculture is based on costs and crop/product yield. It is apparent that considerable expense and effort are expended each year to reduce the impact of weedy plants on crop yields. The purpose of this lab exercise is to explore alternative approaches to weed control in an effort to reduce producer dependence on synthetic chemicals for weed control.

Weeds are plants growing where they are not wanted. **Herbaceous weeds** are common along roadsides, in tree plantations, and in agricultural fields. Herbaceous weeds are monocots (such as grasses) or dicots (pigweed, lambs quarters, and strangler vine, for example). The life cycles, physiology, and plant structure of monocots and dicots are different (see Figure 10-5 in Exercise 10). These differences have important implications for weed management; substances that inhibit the growth of monocot weeds should not be used in an area planted in a monocot crop (such as corn, wheat, rye, or sorghum, for example). A compound that actively inhibits dicot seed germination would not affect the germination rate of a monocot crop such as oats. Hardwood trees growing wild in undesirable locations are **woody weeds**, distinguishable from the herbaceous counterparts by the presence of woody tissues. Chemical treatments for the effective control of woody plants are different from those used for herbaceous weeds.

Many attempts at chemical control of weeds involved the use of synthetic plant hormones that stimulate excessive growth beyond the plant's ability to function or inhibit growth, such as seed germination. Some of the earliest synthetic hormones mimicked the action of the naturally occurring plant hormones, gibberellins, auxins, and cytokinins. Prior to its removal from the market by the Environmental Protection Agency in 1979, 2,4,5-T was used extensively in the United States. This herbicide caused massive defoliation of hardwood seedlings in young pine stands and provided exceptional control of grasses in pine stands, right-of-ways, and along roadsides. It was removed from the market as a result of its extreme toxicity; 2,4,5-T has been linked to liver cancer and birth defects in humans with similar effects on other vertebrates such as fish, birds, and mammals. Replacements for 2,4,5-T have been developed, but all of the compounds are toxic at some level. Weed control with these compounds is variable to excellent depending on the application mechanism, time of year, and environmental conditions. A listing of some of the commercially available herbicides is provided in Table 14-2.

One of the most common herbicides in use by the homeowner/small gardener is 2,4-D. When properly applied during a dry period 2,4-D has good success rates in controlling broad-leafed (dicot) herbaceous weeds. This herbicide can be used safely if the application rates

Herbicide		Source
Chemical name	Trade name	
2,4-D	Esteron 99	Dow Chemical
2,4-D	Verton 2D	Dow Chemical
2,4-D + Dicamba	Banvel 520	Velsicol Chemical
2,4-D + Dicamba	Banvel 720	Velsicol Chemical
2,4-D + Dicamba	Trimec 450-E	PBI Gordon
2,4-D + 2,4-DP	Weedone 170	Union Carbide
2,4-DP	Weedone 2,4-DP	Union Carbide
Dicamba	Banvel	Velsicol Chemical
Fosamine	Krenite	DuPont
Glyphosate	Roundup	Monsanto
Hexazinone	Velpar Gridball	DuPont
MSMA	Transvert	Union Carbide
Pidoram	Tordon 10K	Dow Chemical
Pidoram + 2,4-D	Tordon 101	Dow Chemical
Tridopyr (amine)	Garlon 3A	Dow Chemical
Tridopyr (ester)	Garlon 4	Dow Chemical

Table 14-2. Some herbicides approved for use in southern U.S. forests.

and application directions are followed. To test the relative effectiveness of 2,4-D or other appropriate herbicide on a monocot and dicot annual, you will work in groups of two to four students as assigned by your instructor.

EXPERIMENT

Each group of students will be assigned eight field plots or eight greenhouse flats. Plant winter ryegrass seeds (a monocot) in four of the plots/flats and alfalfa or clover seeds (a dicot) in the remaining four plots/flats. If you are using greenhouse flats, be certain that you use the same bag of potting soil to eliminate differences that could be due to soil type. If you are using field plots, use plots which are next to each other in the same part of the field separated by a 2- to 3-foot-wide untilled strip. While it is impossible to assume that the soil conditions in all parts of the field are identical, soil conditions should be the most similar in adjacent locations. Apply the same amounts of fertilizer and water on the same days to each of the plots/flats.

Your instructor will assign you to one of three experimental groups: *preemergent* treatment, *seedling* treatment, and *mature plant* treatment groups. Follow the directions below for your experimental treatment group.

Preemergent Treatment

1. Fertilize and water your plots/flats after soil preparation and allow to stand for 24 hours.
2. Plant four plots/flats with ryegrass seeds and four plots/flats with clover or alfalfa seeds (use either clover or alfalfa in all four locations). Water well and allow 24 hours for the soil to dry.
3. Mix the 2,4-D or other appropriate herbicide according to the application method as directed by your instructor. Saturate the soil surface of two ryegrass plots/flats and two alfalfa/clover plots/flats with herbicide. The untreated plots/flats will serve as control plots/flats. If you are using field plots/flat, do not water for 4 to 5 days. If you are using greenhouse flats, water from the bottom only using horticultural capillary matting.

Safety Note

Be careful when handling the herbicide; wear gloves and safety goggles. Wash your hands completely when finished with the herbicide. Store or dispose of any excess herbicide in accordance with the directions provided by your instructor.

4. Water and fertilize your plants as suggested by your teacher. Greenhouse flats will need to be watered more frequently to prevent the seeds from desiccating during germination. Liquid fertilizer can also be used at the time of watering provided that the application rates are identical for all eight flats/plots.
5. Allow the plants to mature, about 6 to 12 weeks depending on environmental conditions (greenhouse temperatures may accelerate growth).
6. If you are using greenhouse flats, pull up all of the plants from each flat, shake the soil off the roots, and place the plants in a paper bag. Mark the paper bag with your group name, treatment or control, and ryegrass or clover/alfalfa.
7. If you are using field plots, pull up all of the plants in a 1 square foot area from each of the eight plots, shake the soil off the roots, and place the plants in a paper bag. Mark the paper bag with your group name, treatment or control, plot number, and ryegrass or clover/alfalfa.
8. Allow the plants to dry completely (this removes the water) and weigh the contents of your bag. Record the dry weight of your plants in Table 14-3.

Seedling Treatment

1. Fertilize and water your plots/flats after soil preparation and allow to stand for 24 hours.

Treatment Type:
Group Members:

		Dry weight, grams = yield		
Cover plant	**Treatment/control**	**Plot 1**	**Plot 2**	**Average**
Alfalfa/clover	Control			
Alfalfa/clover	Treatment			
Ryegrass	Control			
Ryegrass	Treatment			

Table 14-3. Data for individual treatment group.

2. Plant four plots/flats with ryegrass seeds and four plots/flats with clover or alfalfa seeds (use either clover or alfalfa in all four locations). Water well and allow 24 hours for the soil to dry.
3. Water and fertilize your plants as suggested by your teacher. Greenhouse flats will need to be watered more frequently to prevent the seeds from desiccating during germination. Liquid fertilizer can also be used at the time of watering provided that the application rates are identical for all eight flats/plots.
4. When the seeds have germinated and seedlings are between 1 to 3 inches in height, mix the 2,4-D or other appropriate herbicide according to the application method as directed by your instructor. Saturate the plant surfaces in two ryegrass plots/flats and two alfalfa/clover plots/flats with herbicide. The untreated plots/flats will serve as control plots/flats. If you are using field plots/flats, do not water for 4 to 5 days. If you are using greenhouse flats, water from the bottom only using horticultural capillary matting.

Safety Note

Be careful when handling the herbicide; wear gloves and safety goggles. Wash your hands completely when finished with the herbicide. Store or dispose of any excess herbicide in accordance with the directions provided by your instructor.

5. Continue to water and fertilize your plants as suggested by your teacher. Greenhouse flats will need to be watered more frequently to prevent the seeds from desiccating during germination. Liquid fertilizer can also be used at the time of watering provided that the application rates are identical for all eight flats/plots.
6. Allow the plants to mature, about 6 to 12 weeks depending on environmental conditions (greenhouse temperatures may accelerate growth).
7. If you are using greenhouse flats, pull up all of the plants from each flat, shake the soil off the roots, and place the plants in a paper bag. Mark the paper bag with your group name, treatment or control, and ryegrass or clover/alfalfa.
8. If you are using field plots, pull up all of the plants in a 1 square foot area from each of the eight plots, shake the soil off the roots and place the plants in a paper bag. Mark the paper bag with your group name, treatment or control, plot number, and ryegrass or clover/alfalfa.
9. Allow the plants to dry completely (this removes the water) and weigh the contents of your bag. Record the dry weight of your plants in Table 14-3.

Treatment of Mature Plants

1. Fertilize and water your plots/flats after soil preparation and allow to stand for 24 hours.
2. Plant four plots/flats with ryegrass seeds and four plots/flats with clover or alfalfa seeds (use either clover or alfalfa in all four locations). Water well and allow 24 hours for the soil to dry.
3. Water and fertilize your plants as suggested by your teacher. Greenhouse flats will need to be watered more frequently to prevent the seeds from desiccating during germination. Liquid fertilizer can also be used at the time of watering provided that the application rates are identical for all eight flats/plots.
4. When plants are mature, 6 to 12 weeks, depending on environmental conditions mix the 2,4-D or other appropriate herbicide according to the application method as directed by your instructor. Saturate the plant surfaces in two ryegrass plots/flats and two alfalfa/clover plots/flats with herbicide. The untreated plots/flats will serve as control plots/flats. If you are using field plots/flat, do not water for 4 to 5 days. If you are using greenhouse flats, water from the bottom only using horticultural capillary matting.

Safety Note

Be careful when handling the herbicide; wear gloves and safety goggles. Wash your hands completely when finished with the herbicide. Store or dispose of any excess herbicide in accordance with the directions provided by your instructor.

5. Wait 7 to 10 days and harvest your plants. If you are using greenhouse flats, pull up all of the plants from each flat, shake the soil off the roots, and place the plants in a paper bag. Mark the paper bag with your group name, treatment or control, and ryegrass or clover/alfalfa.
6. If you are using field plots, pull up all of the plants in a 1 square foot area from each of the eight plots, shake the soil off the roots and place the plants in a paper bag. Mark the paper bag with your group name, treatment or control, plot number, and ryegrass or clover/alfalfa.
7. Allow the plants to dry completely (this removes the water) and weigh the contents of your bag. Record the dry weight of your plants in Table 14-3.

Hypothesis

Write a hypothesis about the effects of 2,4-D (or alternate herbicide) on ryegrass and clover or alfalfa as compared with the control plots/flats.

Hypothesis

Write a hypothesis about the effects of 2,4-D (or alternate herbicide) on ryegrass as compared with clover or alfalfa. On which plant do you anticipate 2,4-D or other appropriate herbicide to be more effective? Why?

Results

Collect data from your classmates and summarize your class findings in Tables 14-4 through 14-6 in the areas provided for the three treatment groups. Which treatment type (preemergent [see Table 14-4], seedling [see Table 14-5], or mature plant [see Table 14-6]) gave the best results for suppression of plant growth?

Preemergent treatment with 2,4-D or other appropriate herbicide summary of class data

		Dry weight, grams = yield		
Cover plant	**Treatment/control**	**Plot 1**	**Plot 2**	**Average**
Alfalfa/clover	Control			
Alfalfa/clover	Treatment			
Ryegrass	Control			
Ryegrass	Treatment			

Table 14-4. Summary of class results for preemergent treatment with herbicide for a monocot and a dicot herbaceous plant.

Seedling treatment with 2,4-D or other appropriate herbicide summary of class data				
		Dry weight, grams = yield		
Cover plant	Treatment/control	Plot 1	Plot 2	Average
Alfalfa/clover	Control			
Alfalfa/clover	Treatment			
Ryegrass	Control			
Ryegrass	Treatment			

Table 14-5. Summary of class results for treatment of monocot and dicot seedlings with herbicide.

Treatment of mature plants with 2,4-D or other appropriate herbicide summary of class data				
		Dry weight, grams = yield		
Cover plant	Treatment/control	Plot 1	Plot 2	Average
Alfalfa/clover	Control			
Alfalfa/clover	Treatment			
Ryegrass	Control			
Ryegrass	Treatment			

Table 14-6. Summary of class results for treatment of monocot and dicot mature plants with herbicide.

On which type of plant, monocot or dicot, did the herbicide work best?

Many herbicides such as 2,4-D are synthetic plant hormones. How does its action affect the large-scale use of herbicides in an agricultural setting?

IMPACT OF AGRICULTURAL PRACTICES ON WEED POPULATIONS

Allelopathy refers to the study of biochemical interactions between plants, fungi, soil algae, and soil bacteria. It is well documented that many plant-derived substances have physiological effects that inhibit or stimulate plant growth. Rice (1995) suggests that allelopathic effects need to be separated from competitive effects such as limited water, minerals, food/nutrients, and light. It is difficult to separate the effects of plant biochemicals on plant growth from growth limited by competition, and most studies of the effects of one plant on another are more accurately described as *interference* studies. Studies from Canada have noted an allelopathic effect of alfalfa (*Medicago sativa*) on itself. Alfalfa seeds planted in soils in which alfalfa was grown the previous two seasons had poor germination rates (Webster et al., 1967). Fallow soils from the same region did not inhibit alfalfa seed germination. Similar studies have also noted the inhibitory effect of alfalfa and alfalfa extracts on cotton (*Gossyplum hirsutum*) (Mishustin and Naumova, 1995; Pedersen, 1965). Liquid extracts of corn (*Zea mays*), soybean (*Glycine max*), oats (*Avena sativa*), and hays containing a mixture of grasses reduced corn seed germination (Martin et al., 1990). These experimental results suggest that the impact of crop residues left in the field in low-till and no-till farming on subsequent crop yield may be significant (see Table 14-7). The results also imply that crop rotation may be essential to maintaining good yields. Although some crops may be suppressed by the previous year's planting, especially if a portion of last year's plants are left to die in the field as a result of cold temperatures and/or desiccation, and are incorporated into soil in the spring (Putnam et al., 1965; Rice, 1995).

Just as crop plants may produce chemicals that inhibit the growth of another crop, many crop plants also inhibit weedy plant growth. Use of organic mulches from some crop plants may also inhibit weed growth. Fall companion planting of winter rye or wheat was found to reduce weed growth in Michigan orchards (Putnam et al., 1983). It was also noted that dried wheat or rye straw applied to the orchard in May produced significant grassy (monocot) weed control for 60 days. Rye and wheat plant residues appear to reduce the amount/number of some broadleaf (dicot) weeds such as redroot pigweed (*Amaranthus retroflexus*), common ragweed (*Ambrosia artemisufolia*), and common lambsquarters (*Chenopodium album*) (Shilling, et al., 1985). Sunflower hybrids grown in rotation with other crops also appear to significantly reduce weed densities (Rice, 1995). Producers can take advantage of these allelopathic effects to significantly reduce weed populations. Rice (1995) also

	Annual crop response							
Cover crop	**Cabbage**	**Carrot**	**Corn**	**Cucumber**	**Lettuce**	**Pea**	**Snap bean**	**Tomato**
Barley (spring)	−	+	+	+	−	+	+	−
Barley (winter)	−	+	+	+	−	+	+	−
Corn	−	+	+	+	−	+	+	+
Oats	−	−	+	+	−	+	+	+
Rye	−	+	+	+	−	+	+	+
Sorghum	−	−	−	+	−	+	+	−
Sorghum X Sudan grass	+	−	−	+	−	+	+	−
Wheat	−	+	+	+	−	+	+	+

NOTE:
+ indicates growth/germination equivalent to that observed under no-till practices without a cover crop
− indicates growth/germination, reduced yield less than that observed under no-till practices without a cover crop

Table 14-7. Effects of spring-seeded cover crop residues on selected annual crops under no-till agricultural practices.

indicates that tilling the soil may significantly increase broadleaf weed populations, thus eliminating the allelopathic effects of crop residues.

Scientists have been able to take advantage of some of nature's control mechanisms by identifying species-specific plant pathogens for some weeds. Skeletonweed (*Chondritta juncea*) was effectively controlled in Australia by the introduction of a fungus, *Puccinia chondrillina,* which targeted the skeletonweed. Significant reductions in the skeletonweed populations were observed (Adams, 1988). Commercial development of some specific plant pathogens, such as the Oomycete *Phytophthora palmivora* (trade name: DeVine) for control of the strangler vine (*Morrenia odorata*), has produced desirable alternatives to synthetic chemical herbicides such as 2,4-D. Inoculation of young yellow nutsedge (a weed common in pastures and rangeland) with spores of a rust fungus, *Puccinia canaliculata,* during humid spring weather magnifies the effect of the rust on the nutsedge, providing an effective natural control (Charudattan, 1991). Genetic engineering techniques hold promise for plant-specific weed management by the genetic manipulation of plant pathogens to accentuate their pathogenic characteristics and host specificity (TeBest, 1991).

EXPERIMENT

You and your classmates will design an experiment to test the weed controlling effects of rye, wheat, and/or oat straw. Your instructor will provide you with seeds of a broadleaf (dicot) weedy plant and/or seeds of a grass (monocot) weedy plant. There are several details to consider:

- What type(s) of weedy plant(s) will be used in the experiment?
- What part of the plant straw will be tested? The whole plant? Roots only? Flowers only? Stems only? Or water/alcohol extracts of the plant?
- What will be used as the control for the experiment?
- How will you determine the results of the experiment, that is, how will you measure the suppression of plant growth? Wet plant weight? Dry plant weight? Numbers of plants?
- At what stage in the weed plant's life cycle will you apply the plant residue? When the seeds are planted? When the seeds have germinated? When the plants are mature?
- Where will you conduct the experiment? In the greenhouse? Classroom? Field plot?

What is the purpose of an experimental control? (*Hint:* See Exercise 2.)

Which type(s) of weed plant(s) will be used in your experiment?

What type of crop plant residue will you use in your experiment? Specify the specific plant and the part of the plant and/or extract to be used.

Where will you conduct your experiment?

How many plants, plots, or flats will you use? (*Suggestion:* Look at the experiment you completed testing the effects of herbicide on weeds.)

At what stage in the weed plant's life cycle will you introduce the crop plant residue?

Hypothesis

Write a hypothesis for testing in your experiment. Be specific and include a prediction of expected results in your hypothesis statement.

Conduct your experiment and record your results in the Table 14-8 provided. Modify the table as needed.

Treatment Type: Crop plant used
Group Members:

		Measure of growth suppression/inhibition		
Weed plant	**Treatment/control**	**Plot 1**	**Plot 2**	**Average**
	Control			
	Treatment			
	Control			
	Treatment			

Table 14-8. Data for individual biological control experiment.

Results

Evaluate the results of your group experiment (see Table 14-8). Was your hypothesis supported or refuted?

Are there any sources of experimental error that you did not identify prior to conducting the experiment that may have affected the outcome of your experiment?

What are the implications of your study results to farming in your area?

Suppose that you and your family have decided to start a small organic vegetable farm and employ natural controls for weeds and insects. Given what you now know about the effects of crop residues, tillage practices, and crop rotation on weeds, design a planting scheme that incorporates early spring planting of peas, broccoli, cabbage, carrots, and onions with summer planting of tomato, pepper, cucumber, lettuce, and herb plants. On the next three pages, draw your spring and summer garden plots separately and indicate where you would mulch and what type(s) of mulch you might use. Include a discussion of fall maintenance such as cover crops, organic mulches, and the like.

GLOSSARY OF TERMS

allelopathy: (*attelon*—of each other; *patheia*—suffering) Addition of one or more chemical compounds to the environment to control weeds.

herbaceous weed: A weed that is an annual (1 year growth), biennial (2 years growth), or perennial (3 or more years growth) in growth habit which does not produce woody tissue; herblike.

herbicide: (*herba*—plant; *cide*—to kill) A chemical or biological agent that is used to slow or stop plant growth; may also kill the plant completely.

weed: Any plant that is growing in an undesirable location; most commonly, a herbaceous plant with rapid growth characteristics which outcompetes a crop species; may also be a tree with undesirable growth characteristics in a managed stand of trees.

woody weed: A perennial weed exhibiting secondary growth, producing woody tissue; an unwanted tree, shrub, or a perennial woody vine growing in an undesirable location.

SPRING PLANTING SCHEME

SUMMER PLANTING SCHEME

FALL MAINTENANCE PRACTICES

EXERCISE 15

Biological Control of Some Insect Pests

OBJECTIVES

Upon completion of this lab exercise, you should be able to:

- list the characteristics of insects.
- differentiate spiders, mites, and ticks from insects.
- describe the three types of insect life cycles—gradual metamorphosis, incomplete metamorphosis, and complete metamorphosis—and give examples of insects with each type of life cycle.
- relate the insect life cycle to the agricultural impact of the insect.
- describe the mechanism of action for ecdysone inhibitors and their implication for agricultural use.
- list the advantages/disadvantages of chemical insect control and understand the use of integrated pest management to make decisions about control measures.
- give examples of insect biological control including:
  - use of insect pheromones in insect traps.
  - use of insect parasites/pathogens.
  - use of insect predators.
- test the effectiveness of *Bacillus thuringiensis, Bt,* for the control of tobacco hornworm.
- test the effectiveness of natural insect controls, diatomaceous dust, on soft-bodied insects.

Suggested Reading:
Chapter 15 of *The Science of Agriculture: A Biological Approach,* 4th edition.

INTRODUCTION

Insects belong to the phylum *Arthropoda*. Members of this group are characterized by an exoskeleton that can be shed, and they have "jointed" paired appendages on some of their segments. The division of the arthropod exoskeleton into distinct plates and cylinders is responsible for the diversity of movements observed among arthropods. Insect bodies are divided into three distinct body regions: **head**, **thorax**, and **abdomen** (see Figure 15-1).

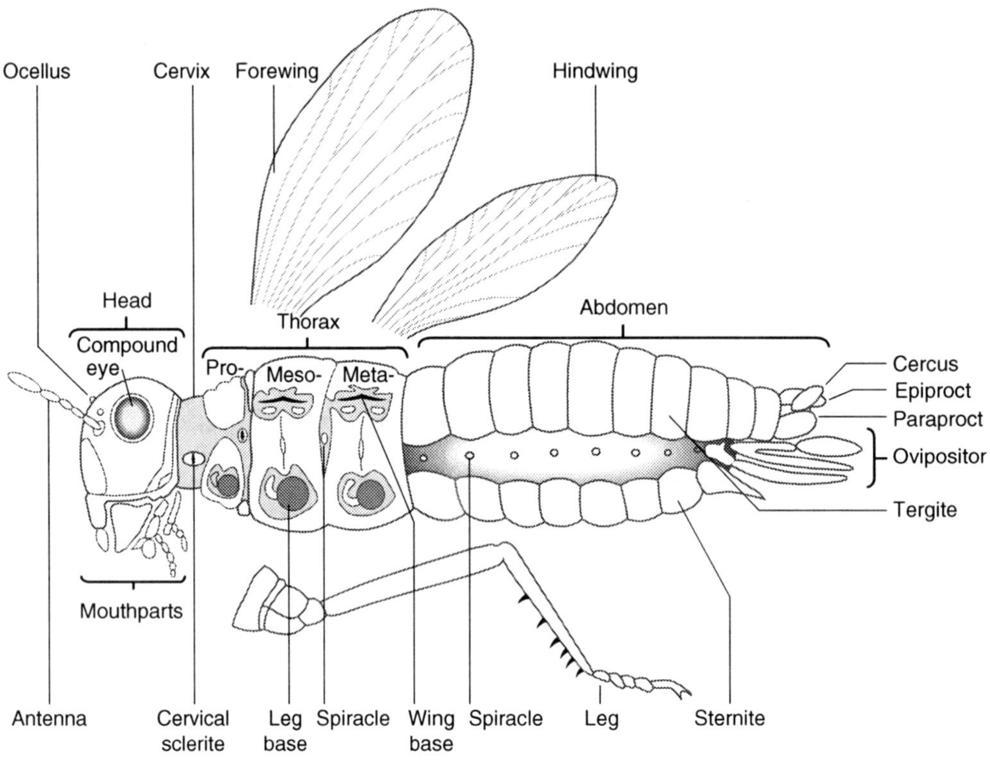

Figure 15-1. The generalized body plan of a winged insect.

In contrast to the annelids, arthropod movement is a segmental process; the internal body cavity is not compressed in the movement process. The number of external appendages is reduced in comparison to the annelids. Each appendage is jointed and highly specialized. Sensory organs are concentrated in the head region and are also highly specialized; receptors that respond to light and chemicals in the environment are common among arthropods. Information about the organism's position and balance is relayed to the brain by specialized receptors, **proprioceptors**. These receptors play a key role in movement and flight. The fact that insects are the only invertebrate group to evolve flight is noteworthy.

The circulatory system is open, that is, blood does not circulate inside closed vessels. Hemocyanin (respiratory pigment) is usually present in the blood increasing the efficiency of oxygen transport. Small respiratory structures, **spiracles** in insects, are responsible for efficient gas exchange. Complex excretory structures such as the Malpighian tubules in insects replace the primitive nephridia common to earthworms. Insects and spiders are able to exploit arid environments more effectively than other invertebrates due to water conservation by the excretory structures and the waterproof cuticle covering the body.

Insects and arachnids (spiders and ticks, for example) are **dioecious**, sexes are usually separate, with internal fertilization following mating. Complex courtship behaviors are observed in many arthropods

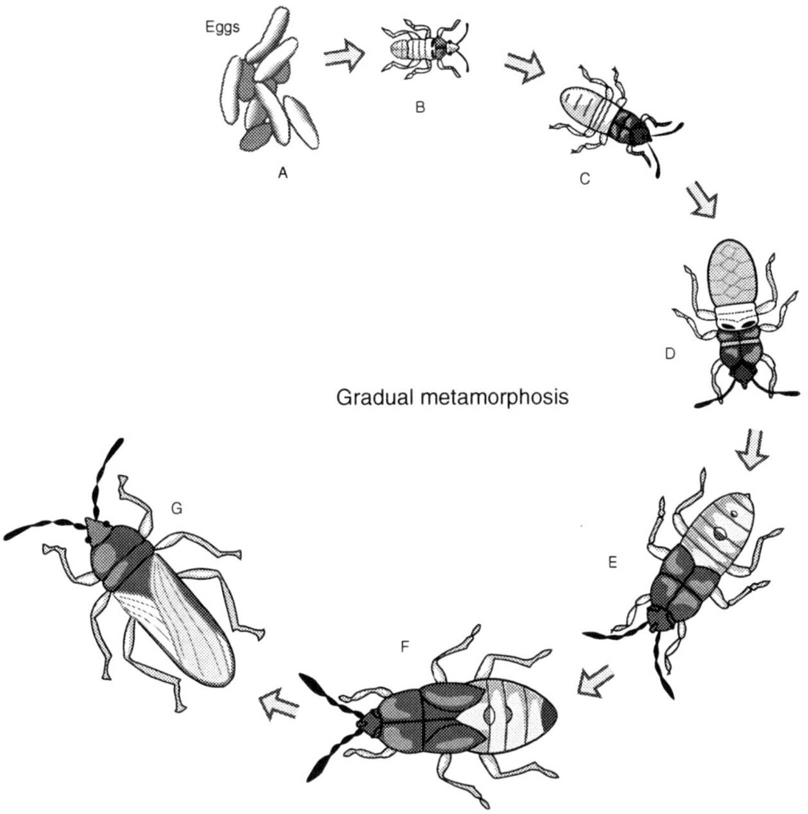

Figure 15-2. Gradual metamorphosis in a cinch bug (Order Hemiptera).

including sexual cannibalism among the praying mantis, some spiders, and desert arachnids. Insect development takes place in a process called **metamorphosis**. Some types of metamorphosis allow the juveniles and adults to live in different habitats, minimizing competition for food and space.

There are three types of metamorphosis: **gradual metamorphosis, incomplete metamorphosis,** and **complete metamorphosis.** Gradual metamorphosis occurs among members of the orders *Orthoptera* (grasshoppers, crickets), *Homoptem* (leafhoppers, cicadas, scale insects, and aphids), and *Hemiptera* (the "true" bugs). Wingless immature insects hatch out of the eggs. The juvenile forms are called **nymphs** and look very much like the adults (see Figure 15-2). Nymphs are usually found in the same environment as adults and must compete with the adults for food. Most insects that undergo gradual metamorphosis feed on plants, and large infestations of these insects during an active reproductive period can cause considerable crop damage in a short period of time.

By contrast, incomplete metamorphosis occurs in insects whose immature forms are aquatic and adult forms are terrestrial. This prevents direct competition between adults and immature forms. Members of the orders *Ephemeroptera* (mayflies), *Odonata* (dragonflies and damselflies), and *Plecoptera* (stone flies) exhibit incomplete metamorphosis.

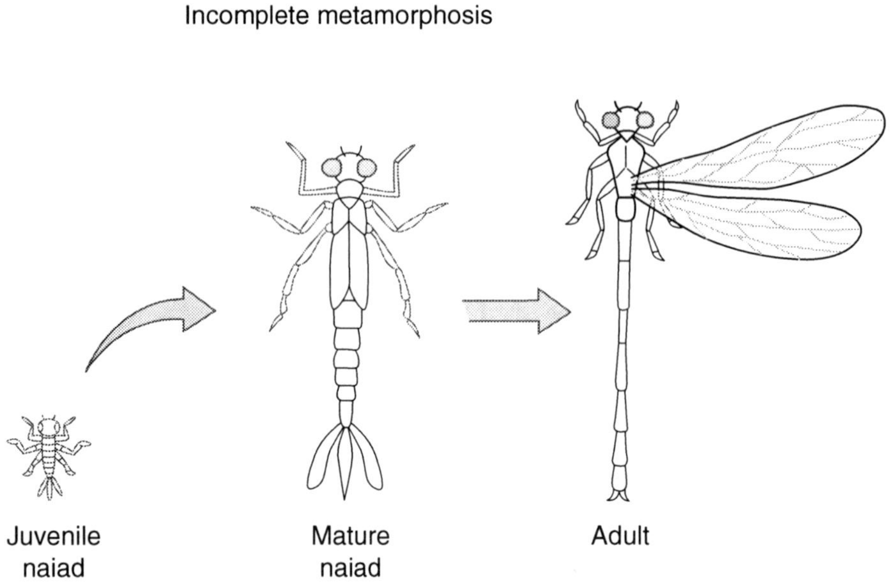

Figure 15-3. Incomplete metamorphosis of the damselfly (Order Odonata).

The wingless aquatic larval form of these insects is called a **naiad** (see Figure 15-3). Naiads play an important role in the food chain as food for many kinds of fish and invertebrates, many spending more than a year in the water before emerging as winged terrestrial adults. The mayfly adult has no mouth; its sole purpose is to mate. After two years as a naiad, the adult female mayfly emerges, mates, and flies over water looking for a place to lay eggs and die. The adults are short lived, giving the group its name, *Ephemeroptera,* meaning *ephemeral* or brief, a direct reference to the short life span of the adult mayfly.

Complete metamorphosis is common to those insects belonging to the orders *Diptera* (flies), *Hymenoptera* (bees, wasps, and ants), *Lepi-doptera* (moths and butterflies), and *Coleoptem* (beetles). The egg hatches into a wormlike soft-bodied juvenile called a **larva** that has specialized mouthparts for eating. The larval forms of many insects have a large impact on agriculture. Some fly larvae are parasites of large animals (such as the warble flies in cattle). Grubs (beetle larvae) cause considerable damage to the roots of plants and to stored grain (white grubs). Some wasp larvae are parasites of other insects, providing natural control for some of the gall-forming insects. After **molting** (shedding of the exoskeleton in response to growth) several times as its size increases, the larva reaches a point where its size is at a maximum, and it molts again to form a cocoon or **pupa**. During the pupal stage, the insect is inactive; it does not eat or move. This is a period of intense reorganization for the insect's body. The winged adult emerging from the pupa case bears little resemblance to the larva (see Figure 15-4). The larvae and the adults of insects undergoing complete metamorphosis rarely share the same habitat or food source, allowing adults to coexist with juveniles at maximum numbers.

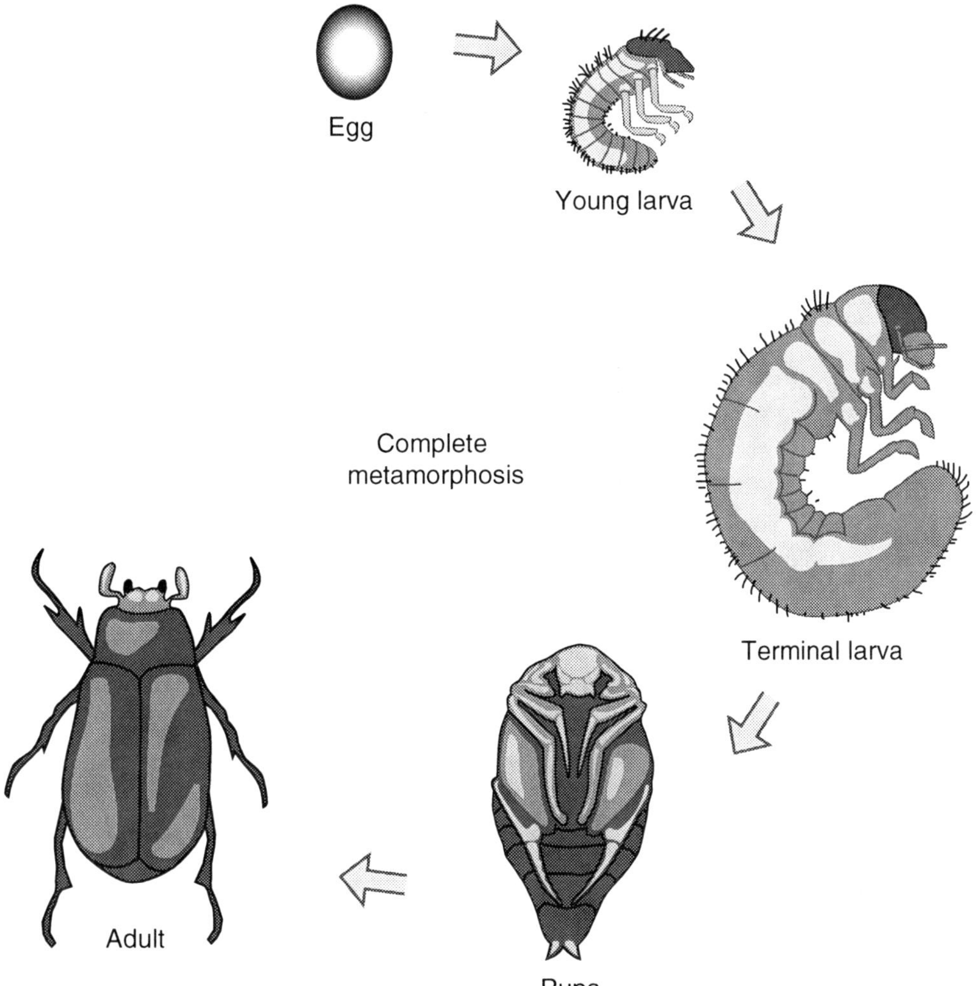

Figure 15-4. Complete metamorphosis in a beetle (Order Coleoptera).

An understanding of the insect pest's life cycle is important to its control. If the life cycle of a parasite is effectively interrupted by the removal of one of its host organisms, a drastic reduction in numbers occurs. Similarly, many juvenile insects are soft-bodied and more easily damaged by abrasive or irritating powders applied to leaf surfaces, like the diatomaceous powder Diapel. During pupation, the insect does not move. Chemicals applied at this stage generally fail due to the protective nature of the pupal case, but the pupae are easily removed by hand if the density is not too high.

DEMONSTRATION

Examine the different insects and the larval forms using a dissecting (stereoscopic) microscope or hand lens provided by your instructor. Determine the type of metamorphosis exhibited by each insect and describe the characteristics of the insect that lead you to your conclusion.

Anthonomus grandis, boll weevil—adult

Gasterophilus intestinalis, botfly—adult

Dung rolling beetle—adult

Manduca quinquemaculata, tobacco hornworm—caterpillar

Tribolium, flour beetle—grub

Hypothesis

Inspect Figure 15-5, the life cycle of the rosy aphid, and identify places in the life cycle where intervention (chemical or natural) would be most effective. Support your hypothesis with evidence about food availability, climatic variability, agricultural practices common to the season, and the like.

Arthropods have the greatest impact on man. Many agriculture pests such as locusts, boll weevils, grasshoppers, aphids, scale insects, and thrips cause significant annual crop losses. Arthropods make significant contributions to disease by serving as vectors for human and domestic animal pathogens (see Table 15-1).

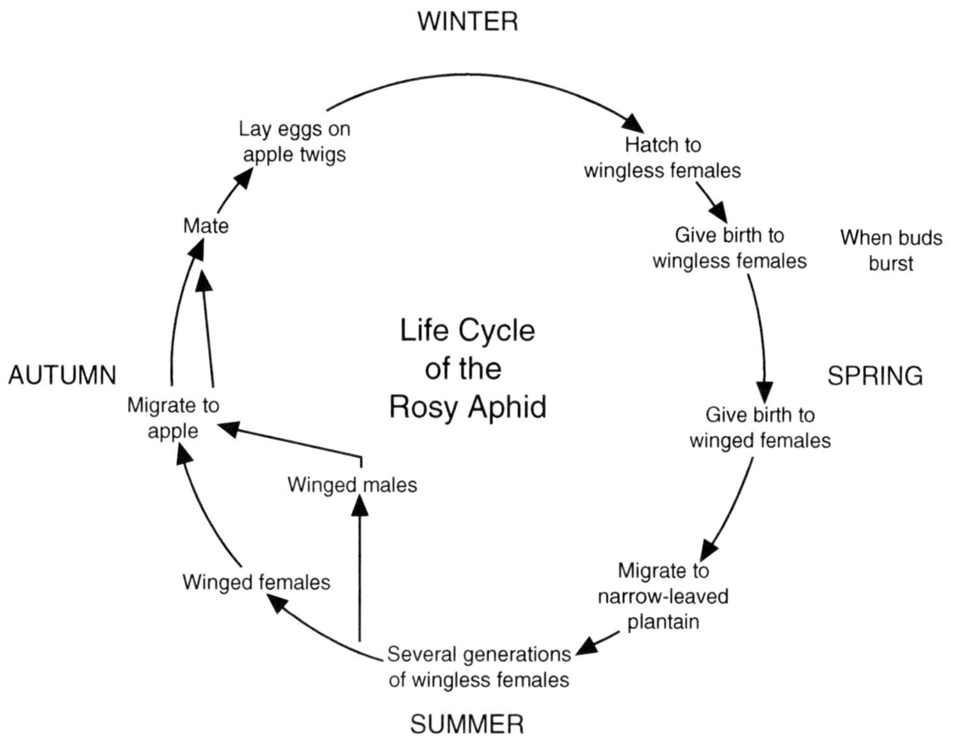

Figure 15-5. The life cycle of the rosy aphid.

Arthropods have changed the course of history. Alexander the Great died of malaria, ending his campaign to expand his empire. Both Oliver Cromwell and James I of England fell to malaria, resulting in a change of English government. During the Middle Ages, bubonic plague reduced the world's population significantly. It is speculated that typhus was responsible for the collapse of the Russian Front in World War I. Similarly, malaria slowed progress on the Panama Canal in the early 20th century. As we began to understand the role of arthropods as agricultural pests and disease vectors, our quest for the control of arthropod pests has been exceeded only by the ability of arthropods to respond to our best efforts. The end result is a new generation of chemically resistant pest organisms. Only biological control efforts coupled with limited chemical application provide any hope of arthropod pest control.

Recent chemical interventions have focused on breaking the insect's life cycle. Some of the most effective chemical treatments for fleas and ants involve the interruption of molting. Since the skeleton of the insect covers its body surface, the insect must periodically shed the exoskeleton when it is too small. A hormone, **ecdysone**, triggers chemical changes in the exoskeleton's surface stimulating molting. One of the first pesticides that effectively inhibited the insect's ability to produce ecdysone was PreCor, approved for the control of fleas. PreCor is extremely effective at interrupting the life cycle; products containing PreCor can guarantee complete control for 60 to 90 days.

Vector	Pathogen	Disease	Host
Anopheles sp (mosquito)	Protozoa	Malaria	Human
Culix sp (mosquito)	Virus	Equine Encephalitis St. Louis Encephalitis	Horse, human
Tsetse fly	Protozoa	Trypanosomiases (African sleeping sickness)	Human
Louse	Rickettsia	Typhus	Human
Culix sp (mosquito)	Filarial nematode	Elephantiasis	Human
Aedes a Anopheles (mosquitoes)	Filarial nematode	Heartworms	Dog
Horsefly	Bacteria	Anthrax	Mammals
Deerfly	Bacteria	Tularemia	Human
Housefly	Bacteria	Cholera	Human
Xenopsylla (flea)	Bacteria	Bubonic plague	Human
Xenopsylla (flea)	Rickettsia	Typhus	Human
Centocephalides felis (flea)	Tapeworm	Intestinal parasite	Cat
Centocephalides canis (flea)	Tapeworm	Intestinal parasite	Dog
Scrub tick	Rickettsia	Scrub typhus	Human
Dermacentor (tick)	Bacteria	Rocky Mountain spotted fever	Human
Ixodes (tick)	Bacteria	Tularemia	Human
Ixodes (tick)	Bacteria	Lyme disease	Human
Dermacentor (tick)	Virus	Hemorrhagic fever	Human

Table 15-1. Arthropod vectors of disease-causing organisms.

Flea eggs hatch, but larvae are unable to molt and die. With the success of PreCor, other insect pests were targeted including the fire ant common to southern states. Improper application of chemicals containing ecdysone inhibitors can adversely affect populations of desirable insects. Remember that insects are the principal pollinators of flowering plants, and a drop in the number of pollinating insects significantly reduces crop yield. Other beneficial insect predators such as praying mantis and ladybugs assist in keeping the numbers of some insect pests low. A decline in their numbers can be potentially disastrous for a grower. Chemicals afford good control of some insect populations. Excessive use of pesticides rapidly leads to the development of pesticide resistance in some insect species. Caution should always be exercised when designing a pest management system; the financial costs should invariably be weighed in conjunction with the environmental costs. The environmental legacy of widespread DDT use during the 1960s necessitates a closer investigation of any pest control program that indiscriminately uses large quantities of chemicals to control insects. Agricultural researchers have learned that an integrated pest management program should contain *direct controls* and *indirect controls*. Direct controls are actions taken by the producer that have an immediate impact on the number of insect pests, such as the application of pesticides. Indirect controls do not have an immediate effect on the number of pest organisms, but they may reduce the population size by inhibiting reproduction, reducing food availability, and/or decreasing the availability of habitats favorable for the pest (see Figure 15-6). Good agricultural practices maximize yield while minimizing negative effects on the environment.

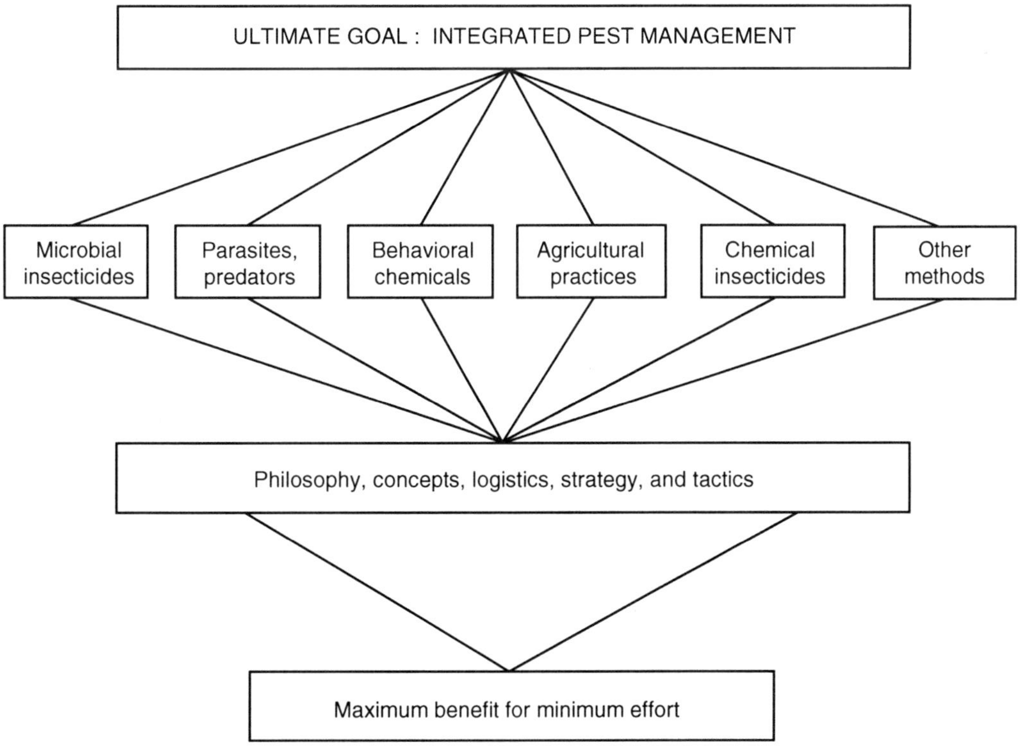

Figure 15-6. Factors affecting decisions in an integrated pest management scheme.

APPLICATION

The spotted cucumber beetle *(Diabrotica undecimpunctata)* and the striped cucumber beetle *(Aclymma vittata)* are disease vectors for *Erwinia tracheiphila,* bacterial wilt of cucurbits (squashes and cucumbers). Carefully analyze the life cycle of bacterial cucurbit wilt as illustrated in Figure 15-7. Use integrated pest management practices to design a control for cucurbit wilt that takes into account the life cycle of the beetle.

How might plant spacing affect the development of this disease?

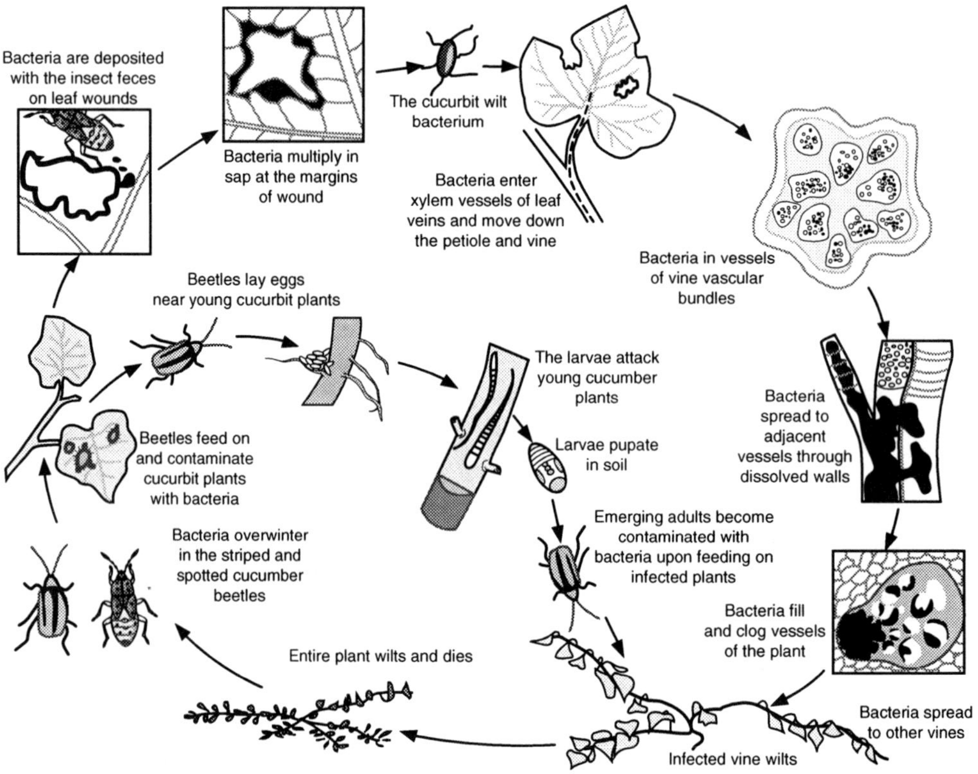

Figure 15-7. Role of spotted and striped cucumber beetles in the life cycle of a bacterial cucurbit wilt.

Considering that most large-scale producers create monocultures of the host organism by planting large fields of one crop, what planting procedures might you propose to limit the beetle's access to its preferred food source (squashes and cucumbers)?

At what point in the life cycle of the beetle do you think the application of chemical pesticides would be most effective?

How important is the management of bacterial wilt of cucurbits to remove diseased plants? Why?

BIOLOGICAL CONTROL OF INSECT PESTS

As of 1992, there were more than 500 insect species that were resistant to at least one synthetic pesticide with some species developing resistance to more than one type of pesticide (Lambert and Peferoen, 1992). In response to increased insect resistance, efforts to develop biopesticides have provided good alternatives to synthetic pesticides. The most successful biopesticide to date is *Bacillus thuringiensis, Bt,* a bacterium that is pathogenic to insects. *B. thuringiensis* produces insecticidal crystal proteins (ICPs) that are highly toxic to some economically important insect pests (armyworms, cotton bollworm, European corn borer, cabbage looper, and the Colorado potato beetle). In 1987, Plant Genetic Systems successfully transferred ICP genes from *B. thuringiensis* to tobacco plants (Lambert and Peferoen, 1992) providing the genetically altered tobacco with natural insect resistance. At the present time, ICP genes have been incorporated into tobacco, cotton, tomato, and potato plants for commercial sale.

Another form of biological control employs **pheromones**, chemical signals produced by insects serving as behavioral stimuli to members of the same species. Ants follow chemical signals laid down by other colony members forming the single file lines of ants all moving along the same trail. The most powerful pheromones for insect control are pheromones involved in reproductive behavior. Some insect traps containing bait with reproductive pheromones are effective in the interruption of the sexual cycle. The southern pine beetle *(Dendroctnus frontalis)* can be lured to sticky traps containing an aggregation pheromone produced by the beetles when they land on host trees.

For small growers, the release of predaceous insects such as ladybugs (for aphids) and praying mantis (for other soft-bodied insects) drastically reduces the population densities of problem insects. Once these predators consume the problem insects, they tend to migrate to other areas where the number of food insects is higher. Natural predators can be effective in the reduction of insect numbers, but they cannot be effective when used alone in a pest management program.

BACILLUS THURINGIENSIS AND THE TOBACCO HORNWORM

You are going to investigate the effectiveness of *Bacillus thuringiensis* in controlling the tobacco hornworm, *Manduca sexta.* Your instructor will divide the class into several groups of three to four students. Each student will receive a tobacco plant and some tobacco hornworm eggs. You will be assigned either the Control group or the *Bt* group. You will be responsible for watering your plants, misting the insect eggs, and hatching larvae. If the eggs dry out completely, they will not hatch. You will also need to water your tobacco plant and provide adequate light during the experiment.

Procedure

1. Mist your plants with water (Control group) or with a solution containing the endospores *of Bacillus fhuringiensis (Bt* group). All of the Control group plants should be sprayed in an area separate from the *Bt* group. Different windowsills or greenhouse shelves/tables should be used for each of the two groups to prevent contamination of the Control group.
2. Thoroughly wet the soil surface of the tobacco plant and allow the leaf surfaces to dry completely before proceeding.
3. Using a camel hairbrush, randomly place the tobacco hornworm eggs on the edges of the tobacco leaves. *Do not share brushes between Control and Bt groups—contamination of the Control group will result!*
4. When all of the hornworm eggs have been positioned on the leaf surfaces, place some wood/plastic plant stakes that are taller than the plant in the soil of the pot. Cover the plant with a loose plastic bag resting on the stakes—this will increase the humidity around the plant, preventing the eggs from desiccating. Do not place your plant in direct sun in a hot window or next to a heat vent.

5. Remember to mist your plants daily and water as needed.
6. Make daily observations about the number of eggs, viable larva, and plant damage for 10 days. Record your observations in Table 15-2 and Table 15-3.
7. Summarize the data for your class in Table 15-4.

Treatment Type: Control or *Bt*										
	Day 1	**Day 2**	**Day 3**	**Day 4**	**Day 5**	**Day 6**	**Day 7**	**Day 8**	**Day 9**	**Day 10**
Number of eggs										
Number of live larvae										

Table 15-2. Individual data for Bt/tobacco hornworm experiment.

Treatment Type: Control or *Bt*											
	Leaf 1	**Leaf 2**	**Leaf 3**	**Leaf 4**	**Leaf 5**	**Leaf 6**	**Leaf 7**	**Leaf 8**	**Leaf 9**	**Leaf 10**	**Average**
Leaf surface area											
Area of leaf eaten by larva											

NOTE:

(1) Multiply total number of leaves by the average leaf surface area = ____ TOTAL SURFACE AREA

(2) Multiply total number of leaves by the average leaf surface area eaten = ____ TOTAL SURFACE AREA EATEN

(3) Divide TOTAL SURFACE AREA EATEN by the TOTAL SURFACE AREA ∞ 100% = ____

PERCENTAGE OF PLANT LEAF SURFACE AREA CONSUMED BY TOBACCO HORNWORM LARVA—record this number to the right: ____%

Table 15-3. Calculations for the amount of plant consumed by the tobacco hornworm.

Treatment	Total Surface Area Eaten	Percentage of Surface Area Consumed
Control group		
Bt group		

Table 15-4. Summary of class data for Bt and tobacco hornworm experiment.

Hypothesis

Write a hypothesis relating the treatment of the tobacco plant with *Bt* spray and insect damage to the tobacco plant leaves.

Results

How long did it take for your eggs to hatch? Did they all hatch?

At the end of the 10-day period, how much did the tobacco hornworm larva affect the leaf surface?

1. To determine the amount of leaf surface area consumed by the caterpillar, trace the outline of the leaf on graph paper divided into 1 inch or 1 millimeter squares. Shade in the area eaten by the insects within the leaf outline. Count the number of squares within the outline of whole leaf to estimate total leaf area.
2. Count the shaded squares to determine the amount of leaf eaten.
3. Repeat this process for one of every five leaves. Average your values to determine the average amount/leaf eaten by the caterpillars.
4. Multiply the average amount consumed by the number of leaves to estimate the total surface area destroyed. Record the values in Table 15-4.

Did the results support or refute your hypothesis?

What possible sources of experimental error can you identify?

Based on your results, do you think that *Bacillus thuringiensis* offers an effective alternative to synthetic chemical controls? Defend your answer using information from your textbook and this experiment.

NATURAL CONTROL OF INSECT PESTS USING DIATOMACEOUS EARTH

Diatoms are unicellular aquatic photosynthetic protists with cell walls of silica. When diatoms die, their cell walls are deposited on the bottom of lakes and oceans. Geologic formations containing large numbers of diatoms are mined for the polishing properties of these fossils. Diatomaceous earth is used in fine abrasives, silver polish, and some grinding papers. In agricultural settings, diatomaceous earth can be used either as a wettable powder or a dust on plant surfaces. Ingestion of the fossils results in massive damage to the insect's gut, and breathing abrasive dust damages insect tracheae. The protective cuticle coating the insect's body surface is scratched by the diatoms, causing substantial fluid loss and dehydration.

EXPERIMENT

You and your classmates will design an experiment to test the effectiveness of diatomaceous earth on an insect pest of your choice. Your instructor will provide you with insect eggs and/or larvae and a host plant. Consider these details:

- What type(s) of insect(s) will be used in the experiment?
- At what stage in the insect's life cycle will you apply the diatomaceous earth?

- How many insects will you use? The same number on each plant?
- How long will you make observations? Days? Weeks?
- Will you use a wettable powder or a dust or both?
- How many times will you apply the diatomaceous earth to the plants in question?
- What will be used as the control for the experiment?
- How will you determine the results of the experiment, that is, how will you measure the effectiveness of the diatomaceous earth? Number of insects killed? Amount of plant growth-dry weight of plants? Amount of leaf/fruit/flower surface area chewed/eaten/ affected?
- Where will you conduct the experiment? In the greenhouse? Classroom? Field plot?

What is the purpose of an experimental control? (*Hint:* See Exercise 2.)

Hypothesis

After discussing the finer points of your experimental design, write a hypothesis that accurately describes your testing of the effectiveness of diatomaceous earth on an agricultural insect pest.

Describe the design of your experiment in the space provided.

Results

Do the results of your experiment support your original hypothesis?

What possible sources of error did you identify as you carried out your experiment?

Based on your personal experience, describe the relative benefits of chemical and biological insect pest control; give specific examples from the experiments that your class has conducted.

What do you identify as the practical implications for large-scale biological control measures in agriculture? (*Hint:* See Figure 15-6.)

GLOSSARY OF TERMS

abdomen: The posterior segment in the insect body; contains reproductive and digestive organs; opening to the reproductive tract, anus, and specialized excretory and respiratory structures.

complete metamorphosis: (*meta*—after, beyond; *morph*—form or shape) Immature insects look very different than the adults; larval forms are generally adapted to different environments/different food sources than the adults thus reducing competition for food and space between the adult and juvenile forms. Many of the agriculturally important insects undergo complete metamorphosis; examples include flies, beetles, wasps, and bees.

$$\text{EGG} \rightarrow \text{LARVA} \rightarrow \text{PUPA} \rightarrow \text{ADULT}$$

dioecious: (*di*—two; *oikos*—house) An organism in which sexes are separate; distinct male or female forms.

ecdysone: (*ekdysis*—getting out; owe—hormone) A hormone produced in the thorax of an arthropod which induces molting and metamorphosis.

gradual metamorphosis: (*meta*—after, beyond; *morph*—form or shape) Immature insects look like miniature wingless versions of the adult forms; juvenile forms of insects that undergo this type of development are called *nymphs*. Examples of agriculturally important insects exhibiting gradual metamorphosis are grasshoppers, thrips, leafhoppers, scale insects, white flies, and mantis.

$$\text{EGG} \rightarrow \text{NYMPH}_1 \rightarrow \text{NYMPH}_2 \rightarrow \rightarrow \rightarrow \text{NYMPH}_n \rightarrow \text{ADULT}$$

head: The anterior segment in the insect body containing simple and compound eyes, mouthparts, mandibles, mouth, antennae, brain, and other sensory receptors.

incomplete metamorphosis: (*meta*—after, beyond; *morph*—form or shape) Insects undergoing incomplete metamorphosis generally have wingless aquatic juveniles, called *naiads,* and terrestrial winged adults; the naiads may bear little resemblance to the adult. Dragonflies, mayflies, stone flies, and damselflies all undergo incomplete metamorphosis.

$$\text{EGG} \rightarrow \text{NAIAD}_1 \rightarrow \text{NAIAD}_2 \rightarrow \rightarrow \rightarrow \text{NAIAD}_n \rightarrow \text{ADULT}$$

larva: (pl. larvae) An immature, wingless, wormlike juvenile insect emerging from eggs; specialized mouthparts allow the larva to feed until it reaches a critical size; then the larva molts and develops into a pupa. Most insect larva have specialized names such as caterpillar (butterfly, moth), grub (beetle), and maggot (fly).

maggot: The larval form of a fly.

metamorphosis: (*meta*—after, beyond; *morph*—form or shape) Process of passing from a larval form to an adult form in the development of an organism.

molt: (*mutare*—to change) The periodic shedding of the exoskeleton (outer covering) or an arthropod in response to growth.

naiad: Immature form of aquatic insects undergoing incomplete metamorphosis.

nymph: Immature form of terrestrial insects undergoing incomplete metamorphosis.

pheromone: (*pherein*—to carry; *mone*—as in hormone) A chemical signal produced by one animal which stimulates behavior in members of the same species.

proprioceptors: (*proprius*—own; *ceptive*—as in receptive) Sensory receptors that allow the insect to respond to internal stimuli; proprioceptors allow insects to properly orient themselves during flight.

pupa: (pl. pupae) A development stage of insects that undergo complete metamorphosis; for example, the wingless larva molts to form a cocoon or case; during pupation, the larval structures are replaced by adult structures. When development is complete, the winged adult emerges from the pupal case.

spiracles: (*spirare*—to breathe) An external opening of the insect tracheal system through which insect gas exchange occurs; usually a series of small openings along the side of the insect thorax and abdomen.

thorax: (*thorac*—breast plate) The middle section of an insect body to which the legs and wings are attached; the thorax of an insect is specialized for the type of movement each insect demonstrates.

EXERCISE 16

Forest Ecology

OBJECTIVES

Upon completion of this lab exercise, you should be able to:

- describe the science of ecology as it applies to forestry.
- discuss the components of the temperate ecosystem including:
  - abiotic components.
  - biotic components.
- distinguish between a community, an ecosystem, and a biome.
- describe the trophic (feeding) relationships in a forest ecosystem.
- take litter samples from a forest floor and identify the trophic levels to which the organisms belong.
- compare biomass estimates for the forest invertebrates collected from a one-square meter sample areas in different forest ecosystems.
- calculate an index of similarity for two forest transects in different locations and relate the forest community structure to the forest practices in the area.
- relate the concept of species diversity to forest management practices.

Note:

This chapter contains concepts and methods in ecology and presumes good knowledge of algebraic functions. A calculator with mathematic functions (exponent, logarithmic) is needed. This exercise is suitable for inclusion in an Advanced Placement Program.

Suggested Reading:

Chapter 16 of *The Science of Agriculture: A Biological Approach*, 4th edition.

INTRODUCTION

Ecology is the study of the interrelationships of organisms with one another and their environment, "the study of organisms at home" (Odum, 1971). Ecology is a complex science incorporating the disciplines of biology, chemistry, physics, geology, geography, and mathematics. Ecological studies can focus on a single organism, a group of organisms living in an area (**community**), or on the organisms and their environment (**ecosystem**). The period between 1830 and 1891 saw increasing exploitation of forested lands in the United States. Wood was a valuable commodity necessary to build railroads to accommodate the westward expansion, to rebuild the South after the American Civil War, and to build houses for settlers in the treeless prairies of the Great Plains. The passage of the Forest Reserve Act in 1891 by the U.S. Congress was a response to the public realization that the forested areas of the United States were not inexhaustible. Beginning with the Yellowstone Park Forest Reservation in 1891, the United States government became the custodian of public forestlands in an attempt to preserve and manage forest resources. Today there are increased demands on forested areas from industrial sources to keep up with the demand for hardwood and paper products. Conservationists are concerned about the potential for large-scale harvesting of old growth forests, particularly in the northwestern U.S. Forest ecologists play an important role by providing information about how to keep a forest healthy to industry, Forest Service officials, and conservation groups. As we begin to move towards sustainable forest management practices, ecologists will continue to provide the scientific information necessary for public policymakers to make wise decisions.

Forest ecology involves the study of the forest ecosystem. An ecosystem is the level of ecological organization encompassing both the living communities of plants and animals and the physical environment in which they live. An ecosystem is composed of two distinct components, the **biotic** and **abiotic** factors. Abiotic factors include the entire nonliving environment such as precipitation patterns, soil type, latitude, altitude, amount of solar radiation, and temperature. Biotic factors are the living portion of the ecosystem, the plants, animals, protists, fungi, and bacteria found in the ecosystem. Management practices affecting the quality of the abiotic portion of the forest ecosystem (such as fertilizers or chemical applications) also affect the living community in the forest. Similarly, removal of several community members also impacts the abiotic segment of the forest ecosystem.

Forests are communities of organisms in which the tree is the dominant vegetation form. Forest ecosystems have high water requirements; the type of tree is determined by the mean ambient temperature and seasonal precipitation patterns. Forests with high rainfall (greater than 80 inches per year) occurring between the Tropic of Cancer and the Tropic of Capricorn are tropical rain forests. Forests in which most of the precipitation occurs in the form of snow with a coniferous dominant tree form occur at high altitudes or northern latitudes in North America. Sandwiched between the tropical rain forest and the northern coniferous forest (sometimes also called the *taiga*) are the temperate deciduous forests. The major vegetation belts (deserts, grasslands, and temperate deciduous forests, for example) existing worldwide are called **biomes**. The vegetation common to each biome is produced in response to specific temperature and moisture patterns of the region. The animal community structure in each biome is related to the complexity of the vegetation community. As the number of plant species within an ecosystem increases, the number of animal, fungal, and bacterial species also increases. The tropical rain forest is the most complex ecosystem on earth hosting the greatest **diversity** of organisms. The biological diversity of an area is the relative abundance of organisms and the number of different kinds of organisms in a community/ecosystem. There are several distinct forest regions in the United States (see Figure 16-1) existing within the four North

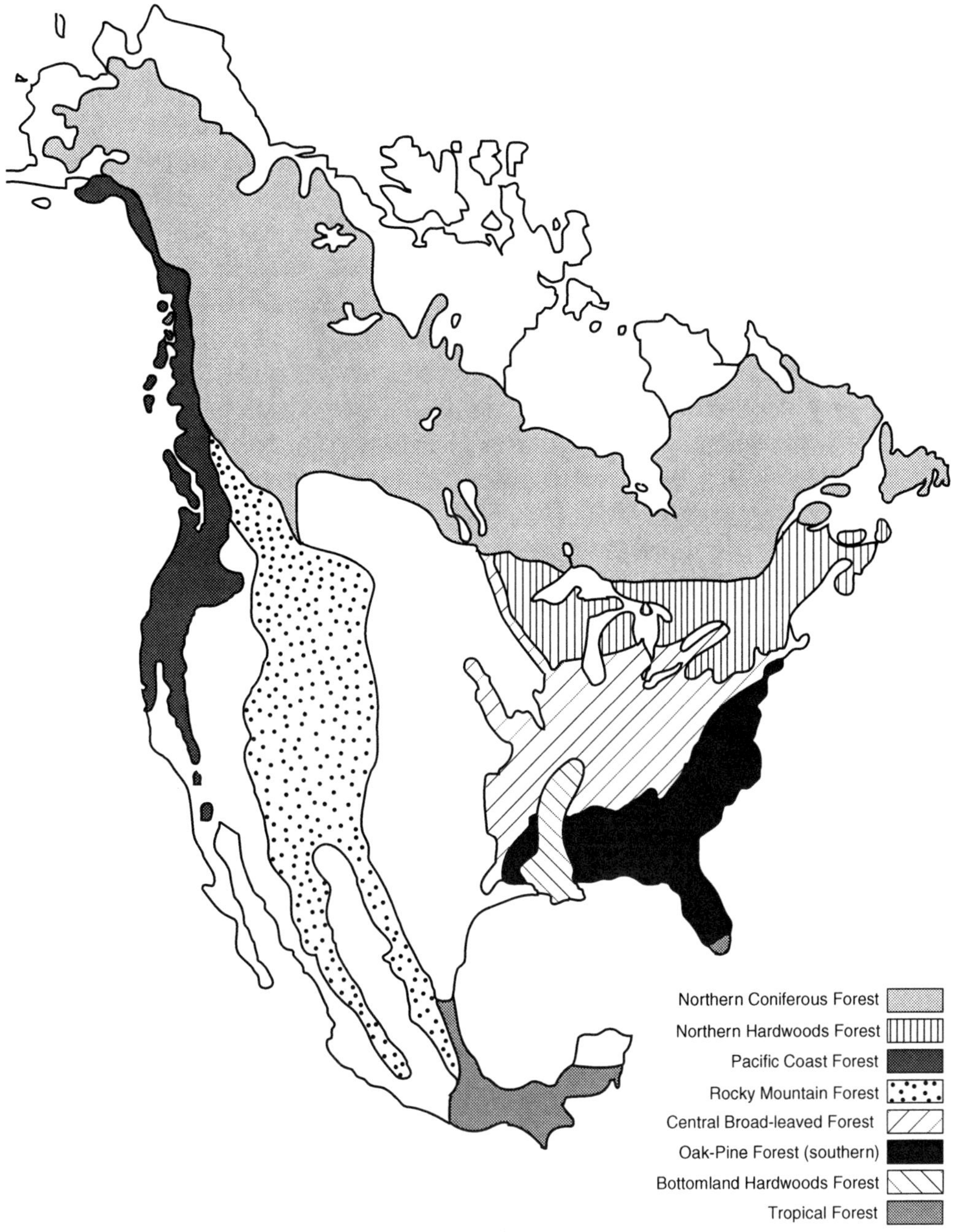

Figure 16-1. North American forest regions.

American forest biomes. Management practices within each of these regions should reflect the differences between the plant and animal communities defining each unique region.

Energy moves through the forest ecosystem in contrast to nutrients, which cycle through the forest ecosystem. The initial source of energy for the forest is sunlight. The forest plants, trees and understory vegetation, use sunlight energy to convert inorganic carbon dioxide (CO_2) and water vapor to organic sugar molecules (glucose) via photosynthesis. Photosynthetic plants and protists are "self-feeders"; that is, they produce their own food from inorganic sources. Self-feeding organisms are **autotrophs**. Autotrophic organisms are called **producers**. In the forest ecosystem, the producers are generally trees, shrubs, and plants growing on the forest floor. Animals feeding on the plants utilize the sugars as an energy source. Organisms feeding on living plant material are called **primary consumers**. The term *consumer* refers to the fact that these organisms cannot manufacture organic carbon-containing molecules from inorganic sources; they must ingest organic molecules from an outside source.

Primary consumers are herbivores, animals that eat plants. Animal predators feeding on the animals that eat plants indirectly use the plant sugars (which have been incorporated into the body of their animal prey) for energy. Animals that eat primary consumers are called **secondary consumers** (see Figure 16-2). Animals that eat secondary consumers are **tertiary consumers**. Organisms that obtain food from outside sources are **heterotrophic**. Energy is transferred from each discrete feeding, or **trophic level**.

Unfortunately, nature is very inefficient. Only 10% of the total energy available at one trophic level is transferred to the next trophic

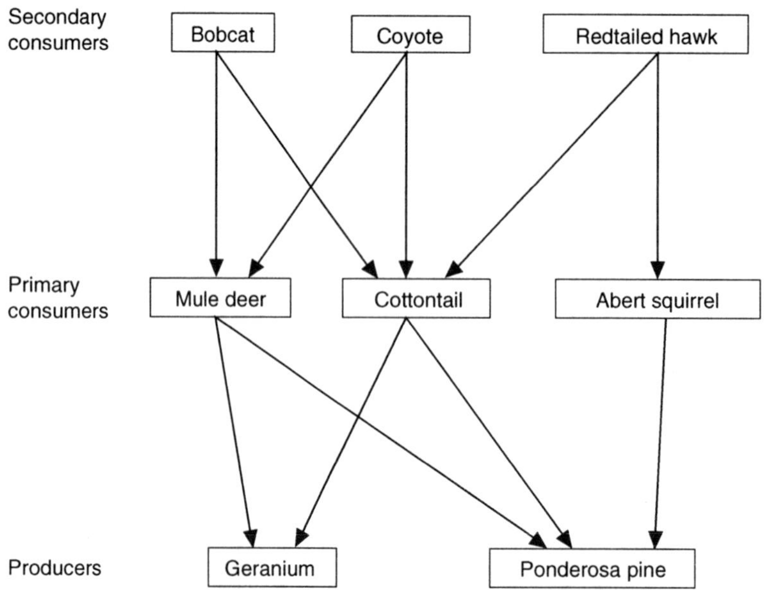

Figure 16-2. A simplified food web with trophic levels.

level. If the primary producers contain 100,000 kilocalories of energy, the primary consumers will receive 10,000 kilocalories. The secondary consumers will have 1,000 kilocalories available to them, and so on. Animals feeding at high trophic levels in a food web must consume large quantities of food in order to obtain enough energy for their metabolic needs. As a result, chemicals present in small quantities at lower trophic levels are concentrated at higher trophic levels in the ecosystem. This is the principle of *biological magnification.* A good example of biological magnification was the accumulation of DDT in bald eagles in the United States during the 1960s. Widespread agricultural and forest application of DDT at low concentrations resulted in high DDT levels in the bald eagle causing thin eggshells. When the eagles attempted to incubate their eggs, the eggshells could not support the weight of the parent bird. As a consequence, bald eagle numbers plummeted and the eagle was placed on the Endangered Species List. Wildlife biologists and ecologists studying the problem correctly identified the source of the problem as DDT. After DDT was removed from the market, bald eagles slowly increased in numbers until the population has almost recovered 30 years later.

Ecosystems must have a mechanism for recycling the dead material, **detritus**, that accumulates. It is essential for the health of the ecosystem that the nutrients and minerals in the decaying material return to the soil or water. Detritivores are organisms feeding on dead, decaying organic matter. Earthworms are examples of detritivores. They ingest leaves and other plant material. As the detritus passes through the gut of the earthworm, it is partially digested and decreases in size. When it is secreted as waste by the worm, the remaining particles are small enough to serve as food for soil bacteria and fungi, the **decomposers**. Decomposers return the inorganic nutrients in the detritus to the soil in a forest ecosystem.

FIELD EXPERIMENT

Your instructor will identify two different forested areas from which samples of the litter on the forest floor are to be collected. These areas should have different management practices, that is, the type and number of different tree species should vary between sites. A commercial pulp (usually pine or some other softwood) or hardwood stand and a natural forest community make ideal study areas. You will work in groups of three to four students to collect, sort, and measure your samples. Your group will collect samples of the leaf/needle litter from several areas in your "study site." When you return back to the classroom, you will separate the live invertebrates from the litter. You will record the number of different kinds of bugs, worms, spiders, millipedes, and other invertebrates in Table 16-1. Examine each type of invertebrate using

Sample 1									
Type of invertebrate									
Location									
Total number									
1. Weight of vial & cap, g									
2. Weight of vial, cap, & live invertebrates, g									
3. Wet weight [(2)−(1)], g									
4. Weight of vial, cap, & dry invertebrates, g									
5. Biomass [(4)−(1)], g									
Trophic level									

Table 16-1. Data for forest litter invertebrate study, Sample 1.

the dissecting (stereo) microscope or a handheld lens and determine the trophic level of each type of invertebrate.

1. Construct a square for collecting which has a total area of one square meter. The simplest way to do this is to nail/glue four meter sticks together in a square.
2. Each student group will need a pair of work gloves for each student, a flat shovel/spade, and large plastic garbage bags with twist ties and labels.

Safety Note

When working in forested areas, remember to wear long pants and shoes that completely cover the feet (ankle boots are preferable in many areas). Always check for ticks when returning from any wooded area. If you decide to use a chemical insect repellent, spray it on your clothes, not your skin. It is advisable to change clothes and shower as soon as you come in from the field if your clothing is saturated with insect repellent.

3. Your teacher will assign each student group to a general location in a study area. To randomly collect a sample, most ecologists use a gridded map of the area and number the grids. Using a random number sequence, a subset of grids is used as sample sites. For our purposes, pick a line on the horizon and walk toward it. At the fifth tree, stop and throw the square meter collection plot in any direction. Collect everything inside of the square, using the shovel to remove all of the leaves, twigs, and other dead vegetation from the soil surface and place it inside of the garbage bag. Quickly tie the bag shut to prevent the escape of any invertebrates. Write the collection date and location on the tag. If you cannot sort your samples immediately when you return to the classroom, the bags can be refrigerated until you have time to work on them.
4. Repeat this process until you have collected three different samples from your designated site.
5. To sort your samples, you will need a metal or enamel rectangular pan, a pair of tweezers (or forceps), and some small glass vials with screw caps. As you sort your samples, weigh each vial with its lid, and put all of the organisms that belong to the same group of invertebrates (all of the beetles, ants, leafhoppers, worms, spiders, etc.) for each sample in one vial. Label the vial cap with the collection date and site location. Weigh each vial and its contents and record the weight (grams) in Table 16-1. Subtract the weight of each vial from the weight of each vial containing invertebrates. This is a measure of the wet weight of the organisms.
6. Allow your invertebrates to dry (in an oven at low heat) until all of the water has been lost.

7. Reweigh the vial and its contents. Subtract the weight of the vial from the weight of the vial with the dried organisms inside. This is a measure of dry weight or **biomass**. Biomass is the "living weight" of the organisms in an ecosystem. Record the dry weights for each group in Table 16-1.
8. Repeat this process for your second and third samples. Record the data for your second sample in Table 16-2 and the data for your third sample in Table 16-3.

Hypothesis

Based on what you know about the management practices in the two study areas selected by your instructor, write a hypothesis relating invertebrate diversity, biomass, and forest type.

Results

1. Average the biomass estimates (g) for each group of organisms from the three sites. For instance, if Site 1 had 10.5 g of beetles, Site 2 yielded 3.0 g of beetles, and Site 3 contained 2.5 g of beetles, the average (mean) beetle biomass for your study area would be (10.5 + 3.0 + 2.5)/3 = 18/3 = 6.0 g. Repeat this process for each invertebrate group you identified. If the group was missing from one of the sites, add the two biomass numbers for the sites, but remember to divide by three. Record your values in Table 16-4.

   Use the equation below if you are having difficulty. Remember to record your values in grams.

$$\frac{[\text{BIOMASS}_1 - \text{SAMPLE}_1 + \text{BIOMASS}_1 - \text{SAMPLE}_2 + \text{BIOMASS}_1 - \text{SAMPLE}_3]}{3}$$

2. The class will need to pool their data for each study site. Record the average biomass values for each invertebrate group for all of the class samples from the two different study sites in Table 16-5. Give a copy of Table 16-4 to your instructor.

Sample 2									
Type of invertebrate									
Location									
Total number									
1. Weight of vial & cap, g									
2. Weight of vial, cap, & live invertebrates, g									
3. Wet weight [(2)−(1)], g									
4. Weight of vial, cap, & dry invertebrates, g									
5. Biomass [(4)−(1)], g									
Trophic level									

Table 16-2. Data for forest litter invertebrate study, Sample 2.

Sample 3									
Type of invertebrate									
Location									
Total number									
1. Weight of vial & cap, g									
2. Weight of vial, cap, & live invertebrates, g									
3. Wet weight [(2)−(1)], g									
4. Weight of vial, cap, & dry invertebrates, g									
5. Biomass [(4)−(1)], g									
Trophic level									

Table 16-3. Data for forest litter invertebrate study, Sample 3.

	Type of invertebrate								
Location									
Biomass—Sample 1									
Biomass—Sample 2									
Biomass—Sample 3									
Mean (average) biomass (g) $\frac{[\text{Sample 1} + \text{Sample 2} + \text{Sample 3}]}{3}$									

Table 16-4. Mean (average) biomass for invertebrate groups in individual forest study site.

	Type of invertebrate								
Biomass numbers by location									
Site 1									
Site 2									

Table 16-5. Class data, biomass estimates for different invertebrate groups in two different forest study sites.

Did your results support your original hypothesis? (Analyze Table 16-5.) Explain your results and discuss any sources of experimental error.

Was there a difference in the numbers of primary, secondary, tertiary consumers, and detritivores among invertebrate groups in the two sites? Which site had the greatest number of detritivores?

Which site had the greatest primary consumer biomass?

Based on your field observations, which site do you think had the highest environmental quality, was the "healthiest" ecosystem? (*Hint:* Look at invertebrate diversity.)

Did this experiment have a control? Explain.

APPLICATION

You work for the U.S. Forest Service as the forester responsible for making decisions about timber harvesting practices in a national forest located in the Appalachians. The national forest contains some old-growth stands of oak, hickory, and chestnut, and some pine stands in drier areas on the south-facing slopes (these stands were recently logged). It also contains hemlock, tulip poplar, oak, beech, hickory, maple, and dogwood in protected moist areas on the north- and east-facing slopes. In which of these communities would you expect to find the greatest animal diversity?

Why?

A furniture manufacturer is lobbying the Forest Service to selectively log the old-growth areas containing poplar, hemlock, and beech. Predict the effects of logging equipment and tree removal on the old-growth forest. Discuss how you would attempt to minimize the impact on the forest ecosystem, as well as address the needs of the logging industry.

A COMPARISON OF TWO FOREST COMMUNITIES: AN INDEX OF SIMILARITY

The index or coefficient of similarity is a measurement of the extent to which two habitats or communities share the same species or individuals (Southwood, 1978). There are several different methods for calculating similarity coefficients, but the method most commonly used by foresters is a modified form of Sorensen's coefficient (Southwood, 1978):

$$C_s = 2jN/(aN + bN)$$

where a = number of species in Habitat a; b = number of species in Habitat b; j = number of species shared by both habitats; and N = number of individuals sampled.

To see how this is calculated, let's look at a practical example.

Species	Maple	Red Oak	Shagbark Hickory	River Birch	Hemlock	
Habitat a	17*	42	28	4	0	91 = Na
Habitat b	22	19*	20*	0	15	76 = Nb

The lowest number for shared species is marked (*)—each habitat has at least that many individuals of the particular species. The marked (*) values are added to arrive at jN (17 + 19 + 20 = 56). The coefficient of similarity would be (2 × 56)/(91 + 76) or 112/167 which is 0.6707. Forest habitats a and b have 67.07% of individuals and species in common. The closer the coefficient value is to one, the greater the similarity between the two forest habitats. If the value is close to zero, the habitats are very dissimilar. The index of similarity is used by foresters to identify forest habitats that are highly similar in plant community structure. Since the plant community structure determines the animal diversity to a large extent, habitats with high indexes of similarity would most probably have very similar animal communities.

You and your classmates will go back to the two different forest sites selected by your instructor. Each group of students will walk a transect line of length determined by your instructor. You will need to wear appropriate clothing, and carry a field notebook, pruning shears, and collection bags. As you walk your transect, you need to count the number of different kinds of trees. You may want to remove a small twig with two to three leaves from each tree on your transect line (you do not want to disturb the site more than necessary, so do not take large limbs). Put your twigs in a collection bag. Record the number of each tree type in your transect line. If you are not certain if two trees are the same, compare the leaf shape, arrangement, and bark pattern on twigs from each tree. You do not have to identify each type of tree to species (sugar maple, red oak, etc.), simply record them as Species 1, Species 2, and so on. Each student group should walk a transect in each area (this can be done on two different days). If you collect twigs, you need to look at them the day they are collected or store them in the refrigerator or a cooler until you have time to analyze your samples.

1. Working in groups of three to four students, analyze the twigs collected from both transect sites. Record the information in Table 16-6.

Species	Habitat a	Habitat b
Species 1		
Species 2		
Species 3		
Species 4		
Species 5		
Species 6		
Species 7		
Species 8		
Species 9		
Species 10		
Species 11		
Species 12		
SUM the columns	aN	bN

* Mark the lowest number which appears in a row. Add all of the marked (*) values to obtain jN = ____ Index of Similarity = ____

Table 16-6. Data table for transects through two different forest sites.

2. Calculate the coefficient (index) of similarity:

$$C_s = 2jN/(aN + bN)$$

where a = number of species in Habitat a; b = number of species in Habitat b; j = number of species shared by both habitats; and N = number of individuals sampled.

How similar were the two forested areas you sampled?

Are any similarities due to forest management practices, or do you feel that they are a result of abiotic factors such as forest age, soil type, and climate? Explain your answer.

Thinking about the data you collected in these two different forested areas, what is the relationship between species diversity and forest management practices?

GLOSSARY OF TERMS

abiotic: (*a*—without; *bios*—life) The nonliving components of the environment such as soil type, temperature, and rainfall patterns, etc.

autotroph: (*auto*—self; *troph*—food) An organism that uses chemical energy (such as sunlight) to make their own food; examples: plants and algae.

biomass: (*bios*—life; *mass*—at weight) The living weight in an ecosystem; dry weight which is usually measured in grams/ounces or kilograms/pounds.

biome: Major collections of terrestrial plants, animals, and microorganisms.

biotic: The living components; all of the organisms living in the ecosystem.

community: Populations of different organisms living together in a common environment (location).

consumer: Heterotrophs are also termed *consumers* due to the fact that they consume energy-containing molecules which are either directly or indirectly by-products of the producers.

decomposer: An organism that digests nutrients from decaying organic matter outside its body; the decomposer releases the remaining nutrients into the surrounding environment.

detritus: Dead, decaying organic matter.

detritus feeder: An organism that feeds on decaying organic matter; examples: earthworms, mites, protozoa, insects, and the like.

diversity: A description of the relative number of different species living in a given area; relatively small numbers of individuals belonging to a large number of species produces a high species diversity index; in contrast, large numbers of individuals belonging to one or two species produce a low species diversity index; the Shannon-Wiener species diversity index is calculated as follows:

$$H = -\sum_{i=1}^{S} (p_i)(\log_e P_i)$$

where H = index of species diversity, S = number of species, and pj = proportion of total sample belonging to its species.

ecology: (*oikos*—place to live) The study of the interrelationships of organisms with one another and their environment.

ecosystem: The level of ecological study/organization which includes all of the living organisms in an area and the physical environment in which they interact; living community and the physical environment.

heterotroph: (*hetero*—different; *troph*—food) Organism that cannot make its own food; it must ingest energy-containing molecules produced by other organisms.

primary consumer: (also called *herbivore*) An organism that eats primary producers (plants/photosynthetic protists, etc.).

producers: Autotrophs are also termed "producers" because they manufacture food for themselves and other organisms.

secondary consumer: An organism that eats primary consumers (also termed *carnivore* or meat-eater).

species: A group of organisms living in the same location that can interbreed *and* produce fertile offspring.

tertiary consumer: An organism that eats carnivores (secondary consumers).

trophic level: (*troph*—feeding) Energy flows through ecosystems from producers to consumers at specific levels called feeding (or trophic) levels.

EXERCISE 17

Aquaculture: Water Chemistry

OBJECTIVES

Upon completion of this lab exercise, you should be able to:

- distinguish between lentic and lotic bodies of water.
- describe the characteristics of zones associated with lakes and ponds including the:
  - epilittoral zone.
  - supralittoral zone.
  - littoral zone (eulittoral, upper, middle, and lower regions).
  - littoriprofundal zone.
  - profundal zone.
  - pelagial zone.
- discuss the seasonal temperature stratification in a lake.
- describe the relationship between soluble carbon dioxide (CO_2) and water pH.
- relate soluble oxygen to temperature and the activities of aquatic organisms.
- relate photosynthetic activity of lake phytoplankton to light penetration.
- discuss the role of the following inorganic nutrients in lentic and lotic systems:
  - nitrogen.
  - phosphorus.
- make and interpret measurements of the physical parameters of a body of water including:
  - surface water temperature in two locations.
  - turbidity.
  - light penetration.
  - specific conductance.
- measure and interpret several chemical parameters of a body of water including:
  - pH.
  - total alkalinity.
  - hardness.
  - dissolved oxygen.
  - nitrogen.
- relate water chemistry parameters to the potential productivity of a body of water.

Safety Note:

Several caustic chemical reagents are used in the assessment of water chemistry parameters. Safety goggles with splashguards should be worn at all times when handling chemical reagents. Check with your instructor regarding the proper disposal of chemical wastes. Do not pour test reagents down the drain without instructions from your instructor. When handling water samples, it is always a good idea to wear gloves and to wash your hands thoroughly with antibacterial soap when you are finished.

Suggested Reading:

Chapter 17 of *The Science of Agriculture: A Biological Approach*, 4th edition.

INTRODUCTION

It is important to understand the factors affecting the relative composition of natural waters to maximize productivity in commercial fishponds. Maintenance of water chemistry parameters within the desirable growth range for commercially important fish species is the key to the production of a high-quality product with good yield. Fish are aquatic vertebrates that occur in both freshwater and saltwater environments. This laboratory exercise focuses on the physical and chemical characteristics of freshwater ponds and lakes. If you live in a coastal area, simple modifications of the procedures for testing chemical and physical parameters of freshwater can be made to test estuarine and marine water samples.

Freshwater systems can be divided into two distinct groups: **lentic** and **lotic** systems. Lentic systems are bodies of water in which little or no mass water movement occurs. Oceans, lakes, and ponds are lentic bodies of water. In contrast, lotic systems are characterized by moving water. Creeks, streams, and rivers are lotic systems. Some commercially important species of fish, such as brook trout and rainbow trout, are grown in artificial lotic conditions, but the majority of fish production occurs in lentic systems.

The architecture of standing bodies of water is related to the water level and the occurrence and distribution of vegetation. For example, plants and algae are important sources of food for fish. They assist in maintaining pH and dissolved oxygen levels in lakes and ponds by removing carbon dioxide (CO_2), and by adding oxygen (O_2) as a result of their photosynthetic activity. It is important to note that algae and plants also use oxygen in the process of photorespiration. Yet in a healthy pond ecosystem, the oxygen added by these organisms support the metabolic needs of all aquatic organisms (including fish).

A pond or lake can be divided into six distinct regions (see Figure 17-1). There are two regions above the water line: the epilittoral and supralittoral zones. The **epilittoral zone** is the zone around the shoreline that is not influenced by spray or wave action. The **supralittoral zone** is the area above the highest seasonal water level affected by spray and wave action. The **littoral zone** is the area of the shore and lake bottom between the highest water level and the zone supporting submerged rooted aquatic vegetation. The littoral zone has two distinct areas, the eulittoral zone and the infralittoral zone. The eulittoral zone includes the area of the shore between the highest and lowest seasonal water levels. Due to changes in water level, only plants that can tolerate periods of exposure and inundation survive in this area. The infralittoral zone is the area that actively supports rooted aquatic plants. The upper littoral zone is defined by the presence of emergent aquatic plants (visible above the water). Plants with floating leaves such as water lilies characterize the middle littoral zone. In the lower littoral zone, only submerged, rooted

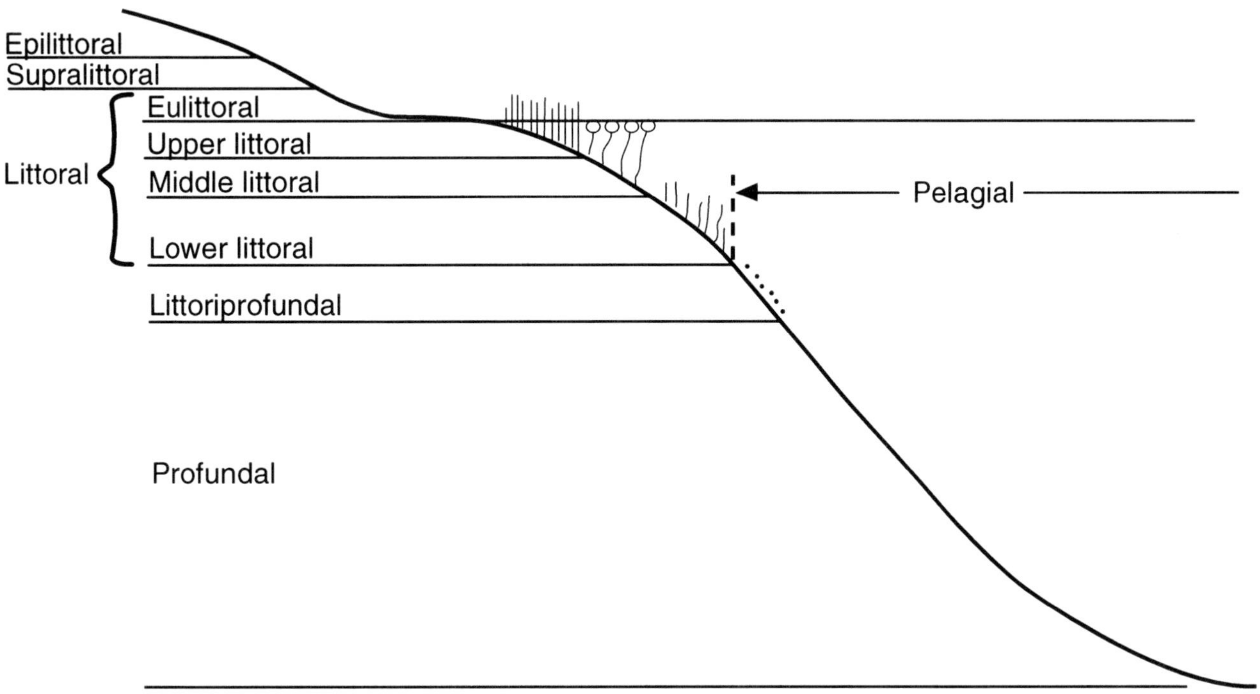

Figure 17-1. Zonation in a standing body of water.

aquatic macrophytes are common. At the point where filtered light levels no longer support the photosynthetic requirements of large plants, algae and bacteria populate the **littoriprofundal zone**. The bottom of the lake, the **profundal zone**, contains sediments devoid of plant life or algae. This is the area where active deposition of lake sediments occurs.

Water temperature affects the chemical and biological components of a lake ecosystem. Oxygen is needed to support the metabolic activities of aquatic organisms. The primary productivity of a body of water is directly related to oxygen availability. As temperature increases, the ability of water to hold oxygen decreases (see Table 17-1). In contrast, as temperature increases, the oxygen consumption rate of fish also increases. The greatest rate of increased oxygen consumption is observed when the water temperature is between 20°C and 30°C (see Figure 17-2).

Oxygen demand is highest in commercial fishponds in the summer when water temperatures are higher and dissolved oxygen concentrations are lower. Aeration of commercial ponds may become necessary in warmer periods. Compounding the problem of reduced oxygen solubility in warmer months is the temperature stratification of the lake or pond occurring in temperate bodies of water. Extremely shallow ponds in which light penetrates to the bottom will thermally stratify. Stratified lakes during summer months will have three different temperature layers with little mixing occurring between regions. The upper layer of water, the **epilimnion**, is warmed by sunlight. Light penetration keeps the productivity high in this portion of the body of water. The cold

Water temperature (°C)	Oxygen (mg/l)	Water temperature (°C)	Oxygen (mg/l)
0	14.621	21	8.915
1	14.216	22	8.743
2	13.829	23	8.578
3	13.460	24	8.418
4	13.107	25	8.263
5	12.770	26	8.113
6	12.447	27	7.968
7	12.139	28	7.827
8	11.843	29	7.691
9	11.559	30	7.558
10	11.228	31	7.430
11	11.027	32	7.305
12	10.777	33	7.183
13	10.537	34	7.065
14	10.306	35	6.949
15	10.084	36	6.837
16	9.870	37	6.727
17	9.665	38	6.620
18	9.467	39	6.515
19	9.276	40	6.412
20	9.092		

Table 17-1. Oxygen solubility as a function of water temperature.

bottom layer, the **hypolimnion**, begins at the depth where light levels are less than 1%. No photosynthetic activity occurs in this region of the lake and the water is uniformly cold. Between the epilimnion and the hypolimnion is a transition zone, the **metalimnion** or **thermocline**, characterized by a pronounced decrease in water temperature of at least 1°C for each meter of depth (see Figure 17-3).

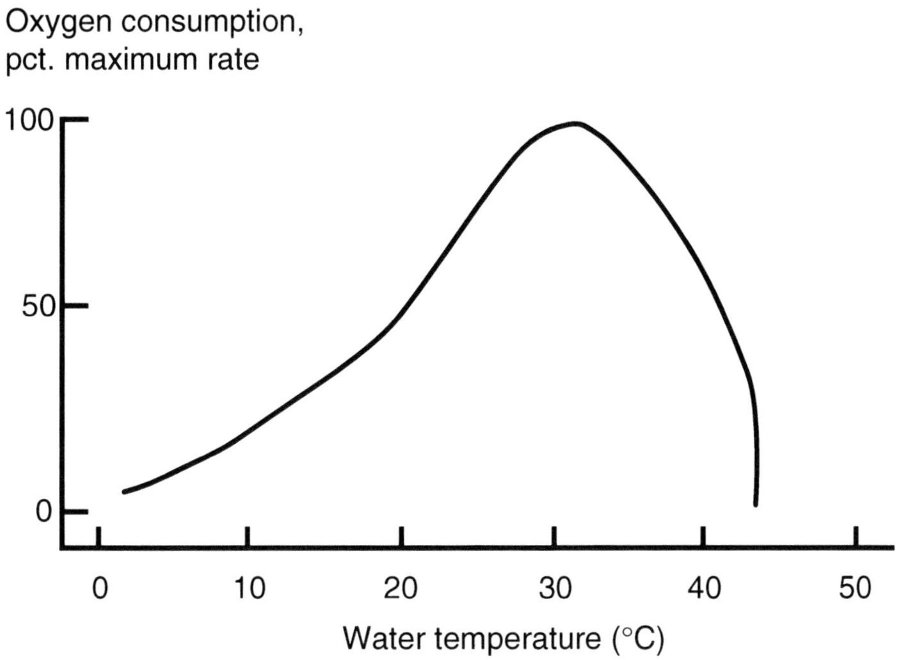

Figure 17-2. Fish oxygen consumption rate as a function of water temperature.

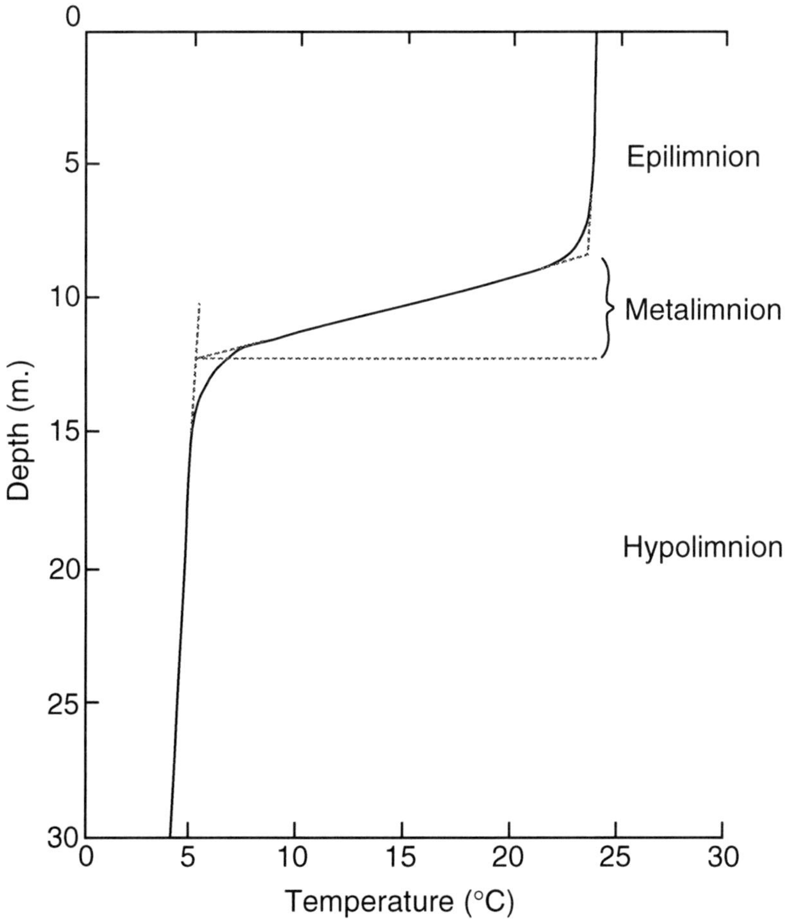

Figure 17-3. Summer thermal stratification in a lake.

Because oxygen solubility is a function of water temperature, dissolved oxygen concentrations should be higher in the hypolimnion. No light filters through to this region, and any biological activity that occurs uses oxygen. Depending on the bacterial activity in the bottom sediments, there may be little or no oxygen available in this region of the lake. Excessive use of chemical fertilizers or fish food stimulates bacterial activity and leads to oxygen depletion in the profundal zone.

Carbon dioxide (CO_2) is produced as a result of heterotrophic metabolic pathways such as cellular respiration. Photorespiration by plants and **phytoplankton** generates the by-product carbon dioxide (CO_2), which is soluble in water. Carbon dioxide (CO_2) dissolved in water (H_2O) forms carbonic acid (H_2CO_3). The form of available soluble carbon dioxide is related to the pH of the pond or lake. **pH** is a function of the number of hydrogen (H^+) ions in solution.

$$pH = -\log [H^+] \quad \textbf{(1)}$$

The pH scale ranges from 1 to 14; 1 is a strong acid and 14 is a strong base (alkaline). Neutral pH is 7.0. Carbon dioxide (CO_2) occurs in four chemical forms in freshwater aquatic systems.

$$CO_2 + H_2O \rightleftarrows H_2CO_3 \quad \textbf{(2)}$$

$$H_2CO \rightleftarrows H^+ + HCO_3^- \quad \textbf{(3)}$$

$$H^+ + HCO_3^- \rightleftarrows 2H^+ + CO_3^{-2} \quad \textbf{(4)}$$

At pH 4.5, the addition of carbon dioxide will not decrease water pH (Boyd, 1990). Since carbonic acid (H_2CO_3) freely dissociates into a hydrogen ion (H^+) and a bicarbonate ion (HCO_3^-) [see equation (3)], most dissolved carbon dioxide exists in the form of the bicarbonate ion (HCO_3^-) (see Figure 17-4).

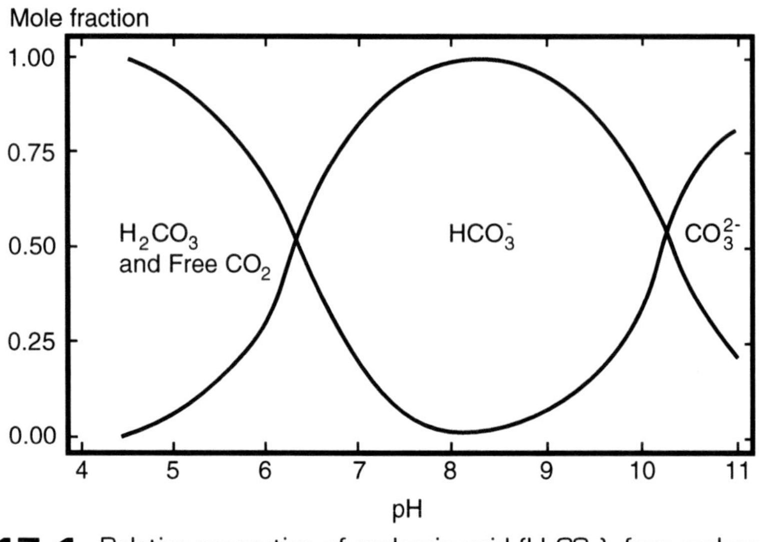

Figure 17-4. Relative proportion of carbonic acid (H_2CO_3), free carbon dioxide (CO_2), bicarbonate ions (HCO_3^-), and carbonate ions ($CO3^{-2}$) as a function of water pH.

The bicarbonate ion further dissociates to form an additional hydrogen ion (H^+) and a carbonate ion (CO_3^{-2}) [refer to equation (4)]. Water with a pH above 8.3 does not contain any free carbon dioxide or carbonic acid; below 8.3, no carbonate ions are formed (Boyd, 1990). Soils with high carbonate and bicarbonate ion concentrations (such as those derived from underlying limestone) tend to produce ponds with slightly alkaline water. These ponds have a greater capacity for buffering the excess carbon dioxide produced by the biological activities of fish. Many of the commercial catfish ponds are located in the southeastern and midwestern states in areas with limestone-derived soils.

Water within the range of pH 6.5 to 9.0 is optimal for fish production (see Figure 17-5). Excessive carbon dioxide production by fish as a result of overcrowding can decrease water pH as low as 4.5. High bacterial numbers related to chemical fertilizer or food input can

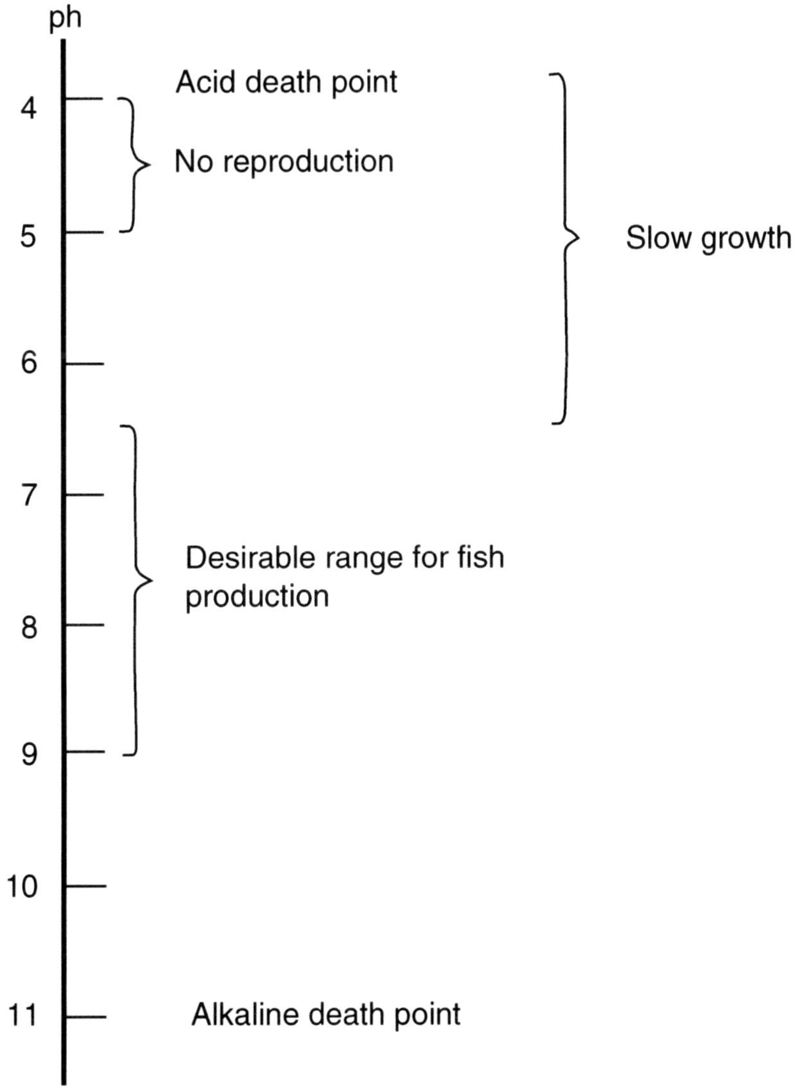

Figure 17-5. Effects of pH on warm-water fish production.

cause a drop in water pH due to increased carbon dioxide production. An algal bloom can produce a drastic drop in water pH below the algal mat in response to increased oxygen utilization by the algae, and an absence of light beneath the algal mat limits oxygen input from the atmosphere and inhibits photosynthetic activity by rooted vegetation. As the fish struggle to get enough oxygen, they tend to deplete available dissolved oxygen and increase dissolved carbon dioxide. This, in turn, decreases water pH. It is not unusual to observe fish kills in conjunction with large algal blooms. There are observable diurnal fluctuations in most ponds. Photosynthetic utilization of carbon dioxide is maximized during hours of peak light penetration, between noon and 6:00 pm. Dissolved oxygen input is greatest during this period, and water pH increases as soluble carbon dioxide decreases. When the sun sets, no photosynthetic activity occurs, dissolved oxygen decreases, soluble carbon dioxide increases, and water pH decreases (see Figure 17-6).

Inorganic nutrients play a role in determining the productivity of an aquatic system. Nitrogen is a nutrient essential to life; proteins and nucleic acids are nitrogen containing organic macromolecules that contain nitrogen. Living organisms need phosphorus for the biosynthesis of nucleic acids, ATP (adenosine triphosphate), and some protein molecules. Nitrogen in aquatic systems exists in several different chemical forms, soluble ammonium ion (NH_4^+), nitrate ion (NO_3^{-2}), and nitrite ion (NO_2^-). Nitrite ion concentrations are usually low in highly oxygenated ponds. Nitrite ions may be produced by bacterial activity in pond sediments if excessive fertilizer application

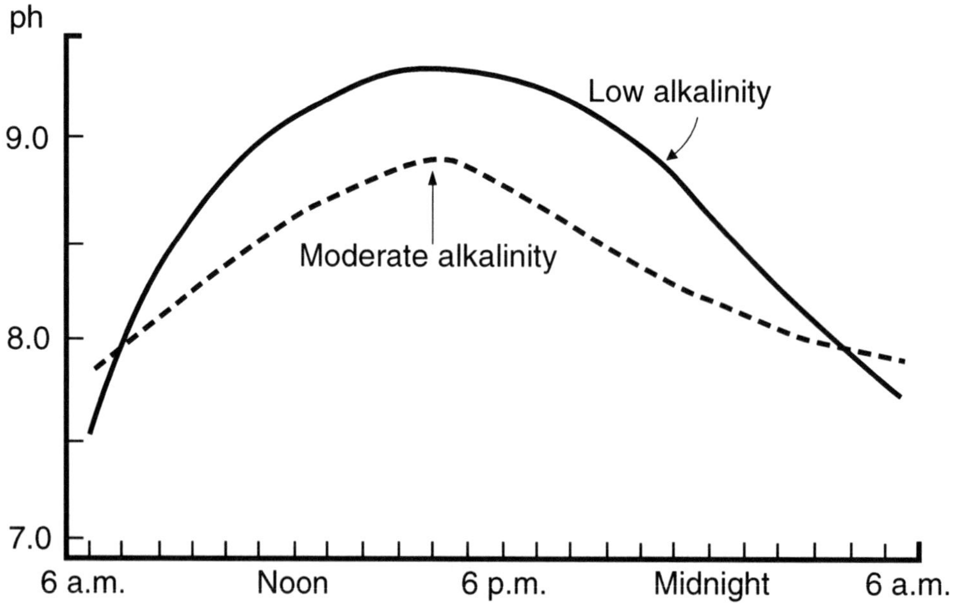

Figure 17-6. Diurnal changes in pH as a function of time.

occurs. The ratio of soluble ammonia to ammonium is a function of water pH; as pH increases, the relative proportion of ammonium ions decreases. The form of soluble nitrogen used by rooted plants is the nitrate ion. Phyto-plankton also use the nitrate ion to support life processes. Without adequate phytoplankton and plant growth, a pond cannot support heterotrophic organisms such as fish and invertebrates.

Phosphorus is the key element in determining the productivity of a body of water. It is generally noted that most aquatic systems will respond to increased phosphorus availability with increased plant productivity (Boyd, 1990). The chemical form of phosphorus in a pond or lake is a function of water pH. Phosphorus dissolved in water forms orthophosphoric (H_3PO_4) acid at extremely low pH. Orthophosphoric dissociates several times depending on water pH.

$$H_3PO_4 \rightleftarrows H^+ + H_2PO_4^- \quad \textbf{(5)}$$

$$H^+ + H_2PO_4^- \rightleftarrows 2H^+ + HPO_4^{-2} \quad \textbf{(6)}$$

$$2H^+ + HPO_4^{-2} \rightleftarrows 3H^+ + PO_4^{-3} \quad \textbf{(7)}$$

As the pH of an aqueous solution of orthophosphoric acid increases, the relative proportions of the four forms change. A pH below 2.0 will favor H_3PO_4; at a pH of 4.75, orthophosphate occurs as $H_2PO_4^-$. At a pH of 9.78, orthophosphate exists as HPO_4^{-2}. Above pH 12.36, orthophosphate is in the form of PO_4^{-3} (see Figure 17-7).

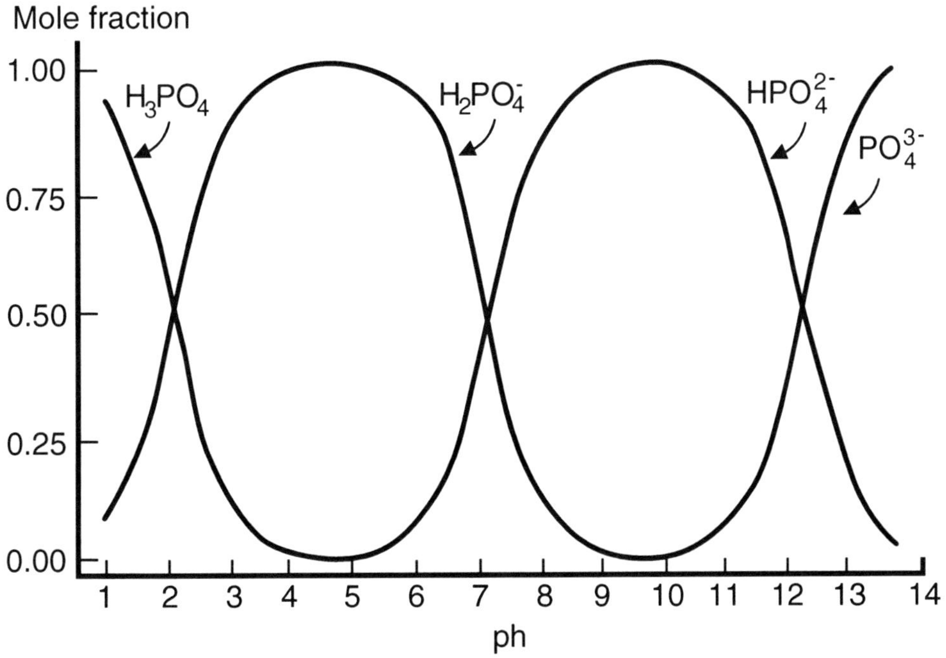

Figure 17-7. Relative concentrations of orthophosphate ions as function of water pH.

Physical parameter	Pond Site 1 (with chemical input)	Pond Site 2 (natural)
Surface water Temperature (°C)	Sunny area:___ Shaded area:___	Sunny area:___ Shaded area:___
Turbidity (JTU)		
Light penetration (meters)	A:___ B: ___ Average (A + B)/2 = ___	A:___ B: ___ Average (A + B)/2 = ___
Specific conductance (μmhos cm^{-1})		

Table 17-2. Measurement of physical parameters for two different pond sites.

MEASUREMENT OF PHYSICAL WATER PARAMETERS

In this portion of the laboratory exercise, you will make measurements on two different bodies of water selected by your instructor. One of the ponds should have some additional source of nutrients such as fertilizer, pasture runoff, or fish food. The other water source should be a natural body of water that is not actively managed for fish production. Some of these measurements can be made on water samples from the sites transported to the laboratory classroom. Other measurements must be made on-site and will require a field trip. Remember to carefully follow the directions given to you by your instructor to prevent accidental injury. You will make several physical measurements on the two bodies of water: surface water temperature, turbidity light penetration, and specific conductance. Record the results of your observations in Table 17-2.

DESCRIPTION OF COLLECTION SITES

Record important information that describes the two sites chosen by your instructor. Include information about the size of the pond or lake, maximum depth (if known), the type(s) of vegetation surrounding the water, and any information about land use in areas adjacent to the pond or lake.

Hypothesis

Write a hypothesis relating the physical characteristics of your two lake/pond sites to active management practices (or lack thereof). Include any pertinent physical information provided by your instructor such as pond area, depth, or location in your hypothesis.

TEMPERATURE

An alcohol or submersible digital thermometer can be used to make surface water temperature readings. Tie a string securely to one end of the thermometer and immerse the thermometer just below the water's surface in a sunny area. Allow the thermometer to equilibrate by waiting 3 to 5 minutes before removing the thermometer from the water. After equilibration, remove the thermometer from the water and read the temperature (°C). Record the value in Table 17-2. Repeat the process for a shaded area and record the temperature value (°C) in Table 17-2.

TURBIDITY

Substances suspended in the water reduce the ability of light to penetrate the water column. **Turbidity** is a measurement of light penetration as a function of suspended solid particles in the water column. High turbidity decreases the productivity of a pond by decreasing light available for photosynthetic activity. Soil particles, inorganic nutrients from runoff, phytoplankton, and **zooplankton** all increase turbidity. You will use a spectrophotometer to estimate turbidity from transmittance readings made in the lab on two water samples.

1. Use demineralized water to zero the Spectronic-20 containing a blue sensitive photocell (see Figure 17-8 for directions on the use of a Spectronic-20).
2. Shake the sample bottle so that the suspended particles are distributed throughout the sample bottle.
3. Fill the 1-inch spectrophotometer tube to the top with the sample water. Place the spectrophotometer tube in the Spectronic-20 so that the white vertical line faces outward.
4. Read the percent transmittance.
5. Find the turbidity units in Table 17-3.

If you have a transmittance of 92%, you would read 90 on the left-hand scale and 2 on the top scale for a turbidity value of 30 JTU. The

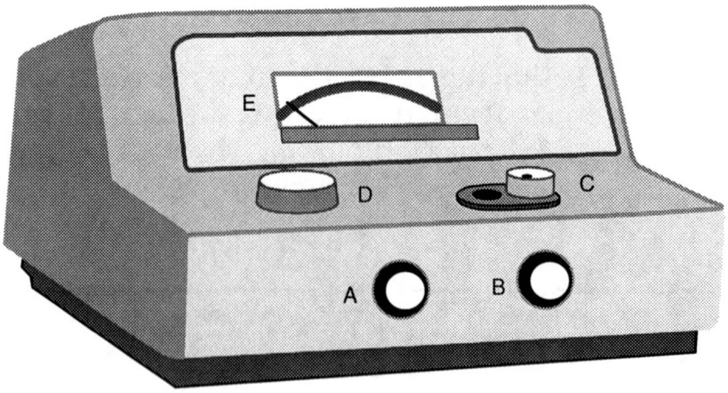

To operate the Spectronic 20 proceed as follows:

a. Turn on the instrument with the amplifier control knob (A) and allow 10 minutes for the instrument to warm up.

b. Turn the wavelength control knob (C) until 450 nm appears in the window next to the wavelength control knob.

c. Zero the instrument by turning the amplifier control knob (A) until the meter needle in (E) reads 0% transmittance. Be sure the sample holder cover (D) is closed when you zero the instrument. Do not zero the instrument with a test tube in the sample holder.

d. Place the blank (Tube #1 containing demineralized) into the sample holder and close the cover. Adjust the instrument to zero O.D. (absorbance) or 100% transmittance by turning the light control knob (B).

e. Remove the blank tube. Insert your sample tube into the sample holder of the spectrophotometer. Make your reading from the transmittance scale on the meter (E). Note that this scale is logarithmic.

f. Repeat steps d and e, to make further measurements. If you change the wavelength control knob (C), you must zero the instrument again.

Figure 17-8. Spectronic 20 for use in estimating turbidity.

Meter readings	0	1	2	3	4	5	6	7	8	9
10	395	380	360	350	335	320	310	300	290	280
20	273	265	258	250	245	240	233	228	222	217
30	211	206	200	197	193	188	184	180	175	172
40	168	164	160	157	153	150	147	144	140	137
50	134	131	128	125	123	120	117	114	112	109
60	106	104	101	99	97	95	92	90	88	86
70	84	81	80	77	75	73	71	68	65	64
80	61	59	56	54	51	49	47	44	42	39
90	36	32	30	26	22	20	16	12	8	4

Table 17-3. Jackson turbidity units (JTU), 450 nm, 1-inch spectrophotometer tubes.

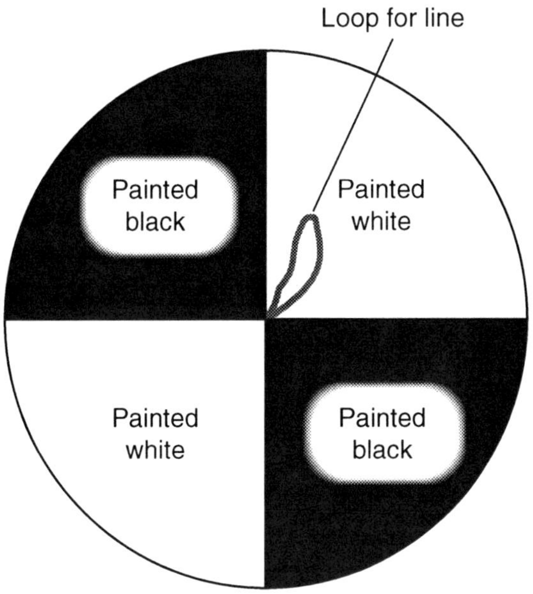

Figure 17-9. Secchi disk.

higher the transmittance value, the greater the amount of light passing through the sample, indicating a lesser number of suspended solid particles. Record your values for the two pond samples in Table 17-2.

LIGHT PENETRATION

A Secchi disk (see Figure 17-9) can be used to determine the depth to which light is visible in the water. The Secchi disk is a weighted 20 cm diameter disk with white and black alternating quadrants. The Secchi disk is attached at the center of the disk to a line with calibrations at regular intervals to measure depth (usually in meters). The same person should make two observations at each pond and take the average of the two readings. This will reduce measurement error by standardizing the readings in both site locations. Record your results in Table 17-2.

SPECIFIC CONDUCTANCE

Specific conductance is a measurement of the resistance of an aqueous solution to electrical current. Specific conductance is important because it relates to the total concentration of dissolved ions. Dissolved ions are one component of lake or pond fertility.

1. You will use a conductivity meter that has been calibrated to a 0.00702 M KCl (potassium chloride) solution at 25°C.

2. Fill two beakers with the water to a point where the vent holes in the dip cell are covered when immersed. Place the two beakers in a water bath and allow them to come to an equilibrium temperature.
3. Rinse the dip cell in the first beaker and transfer the dip cell to the second beaker.
4. Turn on the instrument and gently move the dip cell while keeping the holes completely submerged. Read specific conductance of the scale (in umhos cm^{-1}) and record the value for the sample in Table 17-2.
5. Rinse the dip cell with distilled water and repeat the process with the second sample.
6. When you are finished, turn off the conductivity meter, disconnect and remove the dip cell, rinse the dip cell with distilled water, and allow it to air dry completely.

Results

Did your results support your original hypothesis? Explain any sources of error that might have affected the results.

Which of the two sites is more productive? Include plant, animal, and microscopic organisms in your consideration. Which observations suggest higher productivity?

Which of the two pond sites is best suited for fish production? Why?

MEASUREMENT OF CHEMICAL PARAMETERS

You will make estimates of water chemistry parameters using field test kits such as a Hach or LaMotte kit, on two different bodies of water selected by your instructor. One of the ponds should have some additional source of nutrients such as fertilizer, pasture runoff, or fish food. The other water source should be a natural body of water that is not actively managed for fish production.

Hypothesis

Write a hypothesis that relates the location of the water sample (littoral and **pelagial zones**) to pH, total alkalinity hardness, dissolved oxygen, and available nitrogen.

pH

As mentioned, pH affects many water chemistry parameters such as the relative proportions of various soluble ions. Measurements of water pH should be completed on-site. The most accurate pH measurements are made with a field pH probe. pH indicator paper can be used, but it is not extremely accurate. If you are using a Hach or LaMotte water chemistry kit, colorimetric methods for estimating water pH are included in the test kits. Using pH paper, a water test kit, or a pH probe, measure the pH of the water in an area of the pond near the shore where vegetation is present and in an area in the pelagial zone. Record your values in Table 17-4. Directions for use of a pH meter follow.

1. Obtain 25 ml of buffer solution (which is close to the expected pH value of the water samples) from your instructor and pour it into a small beaker.
2. Open the vent of the reservoir of the electrode located in the plug or sleeve.

Chemical parameter	Pond Site 1 (with chemical input) Littoral zone	Pelagial zone	Pond Site 2 (natural) Littoral zone	Pelagial zone
pH				
Total alkalinity (mg $CaCO_3$/l)				
Free CO_2 (mg/l)				
Hardness				
Dissolved oxygen (mg/l)				
Ammonia (mg/l NH_4^+)				
Nitrate (mg/l)				
Nitrite (mg/l)				

Table 17-4. Water chemistry parameters for littoral and pelagial zones of two ponds.

3. Rinse the electrodes with distilled water and blot dry with soft paper or Kim Wipe.
4. Immerse the electrodes in the buffer solution, measure the temperature of the buffer solution, and calibrate the electrode to that temperature. Three different pH buffers are generally used to calibrate the pH probe to measure a wide range of pH values.
5. Set the function switch to pH and calibrate the probe to the pH of the buffer solution. The probe is now calibrated; *switch the meter to standby.*
6. Rinse the electrodes in distilled water and blot dry.
7. Immerse the electrodes in the sample and measure the temperature of the sample. Set the compensator to this temperature.
8. Set the function switch to pH and gently swirl the sample solution around the probe. As the meter comes to equilibrium, some drift will naturally occur. Read the pH from the scale on the meter and record the value in Table 17-4.
9. Rinse the probe with distilled water and blot dry. *Return the function switch to standby.* Repeat the process with the next sample.
10. *Return the function switch to standby.* Rinse the electrodes with distilled water, and dry.

TOTAL ALKALINITY

Alkalinity refers to the acid-neutralizing ability of natural waters (Wetzel and Likens, 1991). Healthy aquatic systems can absorb excess dissolved carbon dioxide resulting from biological activity. Total alkalinity is the sum of the buffering capacities of the hydroxide ion (OH^-), the carbonate ion (CO_3^{-2}), and the bicarbonate ion (HCO_3^-). Carbonate and bicarbonate ions are metabolic by-products in aquatic ecosystems. Free carbon dioxide can be estimated from total alkalinity and pH (see Figure 17-10). Use the water test kit provided by your instructor to estimate total alkalinity in the littoral zone and the pelagial zone of the two pond sites. Record your values for total alkalinity in Table 17-4. Using Figure 17-10, estimate free carbon dioxide (CO_2) from your alkalinity and pH values. Record your estimates of free CO_2 for each location in Table 17-4.

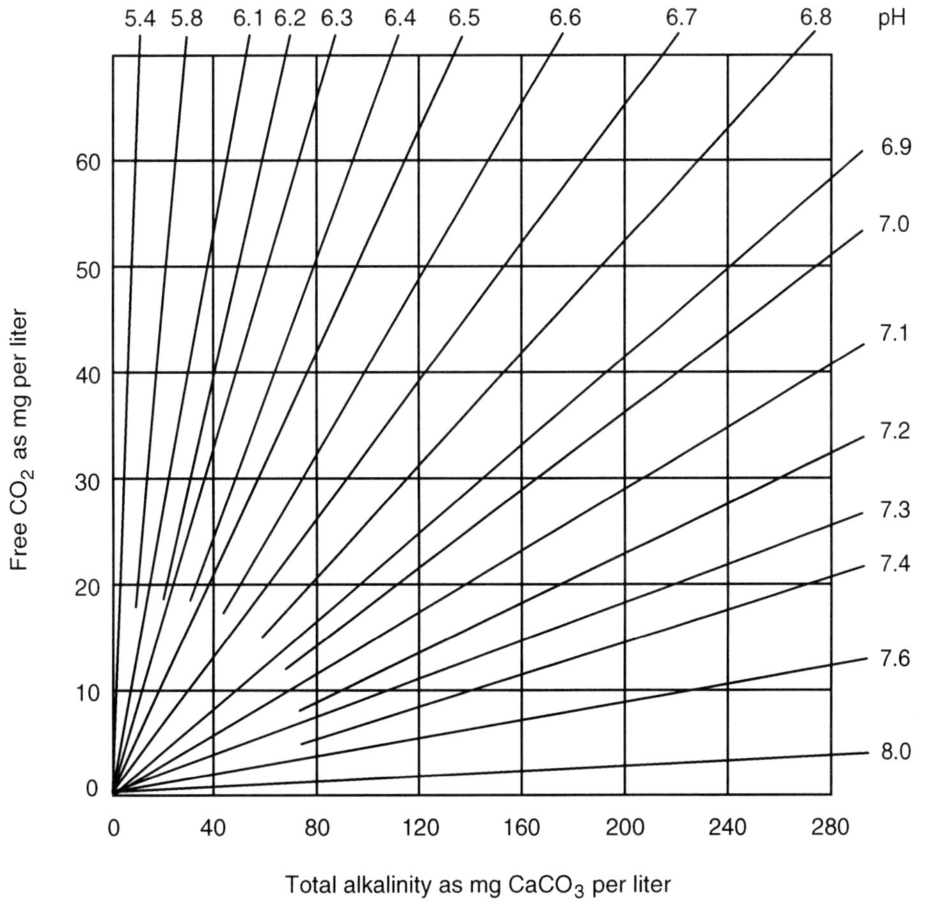

Figure 17-10. Free carbon dioxide (CO_2) as a function of total alkalinity and pH.

HARDNESS

The hardness of water is the ability of polyvalent positively charged ions such as calcium, magnesium, iron, and aluminum, to precipitate soap.

Hardness is a measurement of carbonate and noncarbonate ions. Limestone is the principal ion source for calcium and carbonate ions common in hard water. For our purposes (unless one of your pond sites is on an old mining site), we will consider hardness as a measurement of carbonate ions (mg $CaCO_3$/l). Using the Hach or LaMotte kit supplied by your instructor, measure the water hardness in the pelagial and littoral zones of the two pond sites. Record the values in Table 17-4.

DISSOLVED OXYGEN

Dissolved oxygen (DO) is essential to all aquatic life. Accurate and frequent DO measurements should be made in all aquaculture settings to assure an adequate oxygen supply for the fish. Dissolved oxygen values of 5.0 mg/1 are necessary to support active growth (see Figure 17-11). DO values also manifest diurnal and seasonal fluctuations. All things being equal, dissolved oxygen increases as water temperature decreases. Cold-water fish and fish found in streams with high-stream velocity, such as speckled trout, have high dissolved-oxygen requirements.

Warm-water fish such as tilapia demand less dissolved oxygen than their cold-water counterparts. Dissolved oxygen levels are usually higher near vegetation and DO values peak between noon and 6:00 pm. as a result of peak photosynthetic activity. Dissolved oxygen can be measured with a probe or estimated using the titration methods found in water test kits.

If you are going to use a water test kit, a word of caution is necessary. *The reagents used in estimating dissolved oxygen by titration are caustic and toxic!* Follow the directions in the test kit carefully and dispose of the sample in the manner prescribed by your instructor. Measure the dissolved oxygen in the littoral and pelagial zones of the two pond sites and record the values in Table 17-4. Directions for use of a dissolved oxygen meter follow. You will need a BOD (biological oxygen demand)

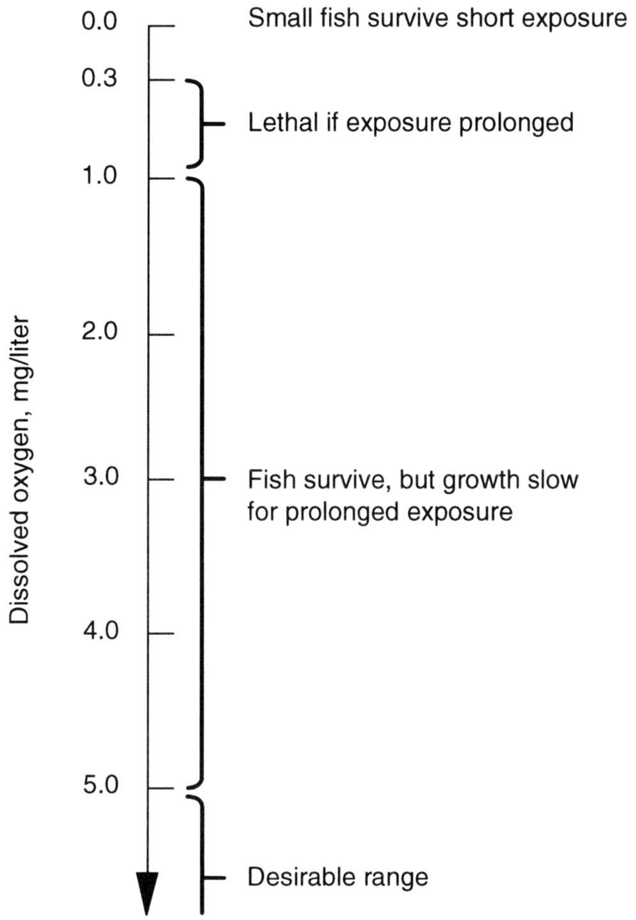

Figure 17-11. Dissolved oxygen (DO) and the growth of warm-water fish.

bottle, a magnetic stirrer, two magnetic stirrer bars, and distilled water for this procedure.

1. Rinse the probe with distilled water and place it in a BOD bottle filed with distilled water. Place a magnetic stirrer bar in the BOD bottle. Turn on the meter and allow it to come to equilibrium.
2. Use a BOD bottle to take a sample. Place a magnetic stirrer bar in the BOD bottle. Place the BOD bottle on a magnetic stirrer and turn the stirrer on, slowly increasing the speed. Take the measurement, allowing 1 to 2 minutes for the reading to stabilize. Record the value in Table 17-4.
3. Rinse the probe with distilled water and return it to the BOD bottle filled with distilled water and allow the probe to come to equilibrium. Repeat the process for the remaining water samples.
4. When you are finished, rinse the probe with distilled water and return it to the BOD bottle containing distilled water. Turn the meter off, but do not allow the probe to dry out.

NITROGEN

Ammonium ions are an important source of nitrogen for bacteria, plants, and algae. Productivity of ponds and lakes is increased by the addition of ammonium nitrate. Nitrate and nitrite ions are additional sources of nitrogen for aquatic systems. The chemical form of the nitrite ion is a function of water pH (see Figure 17-12). Ammonium, nitrate, and nitrite ion concentrations are estimated using colorimetric titration methods and a Hach or LaMotte water chemistry kit. Estimate the concentrations of ammonium, nitrate, and nitrite ions in the littoral and pelagial zones of the two pond sites and record the values in Table 17-4.

Results

Did your results support your original hypothesis? Discuss any unexpected results in the context of the physical characteristics, location, or management practices of the two pond sites.

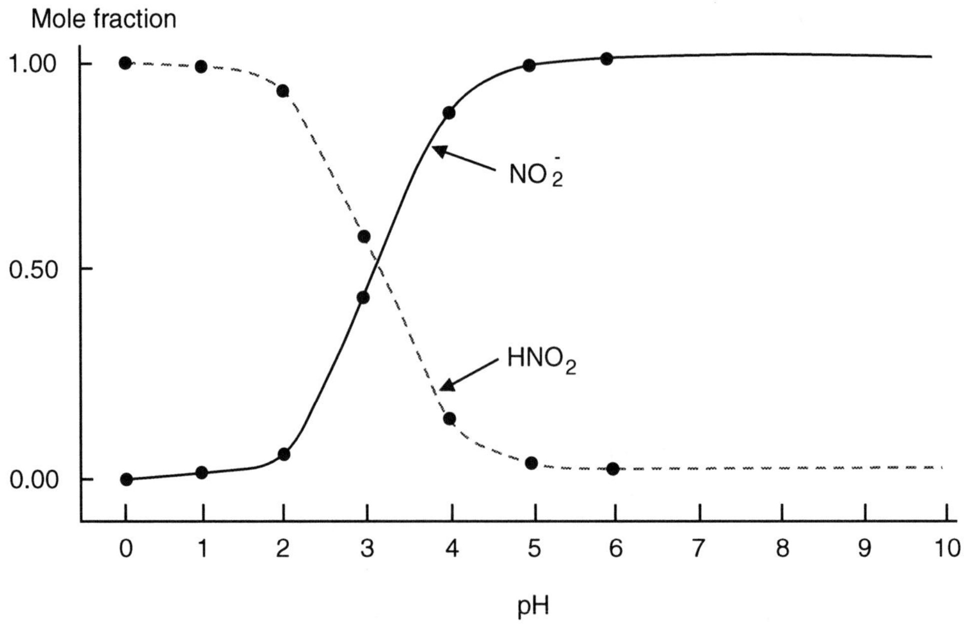

Figure 17-12. Relative proportions of HNO_2 and NO_2~ as a function of water pH.

Which site is most favorable for plant growth (including phytoplankton)?

Which site would be the best location for a commercial fishery? Supply evidence for your decision from the results you collected in Tables 17-2 and 17-4.

GLOSSARY OF TERMS

benthic: (*bathys*— deep) Relating to or occurring at/on the bottom of a body of water.

epilimnion: (*epi*—above; *limne*—lake or pool) The upper layer of warm circulating water above a layer of colder water in a thermally stratified body of water.

epilittoral zone: (*epi*—above; *litordlis*—seashore) The zone above the water line of a pond, lake, or ocean, which is uninfluenced by spray or wave action.

hypolimnion: (*hypo*—below or less than; *limne*—lake or pool) The lower layer of cold circulating water below an upper warm layer in a thermally stratified lake.

lentic: (*lentus*—sluggish) Relating to or living in slow-moving water such as a pond or lake.

littoral zone: (*litomlis*—seashore) The zone of a lake, pond, or ocean between the highest seasonal water level and the area inhabited by submerged rooted aquatic plants. The littoral zone can be subdivided into two distinct regions the *eulittoral zone* and the *infralittoral zone*. The *eulittoral zone* is the area of the shoreline between the highest and lowest seasonal water levels. The *infralittoral zone* is the area of the shoreline colonized by aquatic plants and can be divided into three regions with respect to the type of plants found in each area. The *upper littoral zone* is the area of the shoreline inhabited by emergent aquatic plants. The *middle littoral zone* is the shallow area of the shore in which floating aquatic plants are found, and the *lower littoral zone* is characterized by the presence of submerged rooted aquatic vegetation.

littoriprofundal zone: (*litoralis*—seashore; *profundus*—depth) A transition zone of the shoreline characterized by the absence of rooted aquatic vegetation and the presence of microscopic algae and bacteria.

lotic: living in or related to actively moving water common to streams and rivers.

metalimnion: (*meta*—middle, between; *limne*—lake or pool) The portion of a thermally stratified lake between the upper warm water of the epilimnion and the lower cold water of the hypolimnion; the metalimnion is characterized by a sharp decline in water temperature (thermal gradient); may also be referred to as the *thermocline.*

pelagial zone: (*pelagicus*—sea; more at—lake) The zone of open water in a pond, lake, or ocean.

pH: A measure of the relative concentration of hydronium (H^+) ions. The greater the H^+ concentration, the more acidic a substance is; the lower the concentration of H^+ ions, the more alkaline a substance is.

1		7		14
very acidic	acidic	neutral	alkaline	very alkaline

phytoplankton: (*phyton*—plant; *planktos*—drifting) Microscopic photosynthetic protists (algae) common to the pelagial zone.

plankton: (*planktos*—drifting) Microscopic organisms (usually protists) that swim or drift in the open water of the pelagial zone.

profundal zone: (*profundus*—depth) The portion of the bottom of a body of water devoid of vegetation or microorganisms; characterized by deposits of sediments.

specific conductance: A measure of the resistance of a solution to electrical flow; units of measurement are umhos cm^{-1} (reciprocal of the units of resistance, ohms).

supralittoral zone: (*supra*—above, over; *litoralis*—seashore) The zone entirely above the water level of a pond, lake, or ocean and influenced by spray and wave action.

thermocline: (*thermē*—warm, heat) In a thermally stratified lake, the layer between the upper warm water and the lower cold water in which the water temperature decreases at least 1°C for each meter in depth; also called the *metalimnion.*

turbidity: (*turbidus*—conftised or crowded) The presence of suspended solids (either soil particles or plankton) which reduce light penetration through the scattering or absorbance of light; measured in JTU (Jackson turbidity units) using a spectrophotometer.

zooplankton: (*zoo*—animal; *planktos*—drifting) Microscopic heterotrophic organisms (protists and invertebrates) common to the pelagial zone.

EXERCISE 18

Water Safety

OBJECTIVES

Upon completion of this lab exercise, you should be able to:

- Locate pond organism identification key using the Internet.
- List three different pond organisms found in your sample.
- Identify two different methods of pond organism locomotion.
- Identify organisms as single-celled or multicellular.
- Discuss the importance of water purification systems.

Suggested Reading:

Chapter 18 of *The Science of Agriculture: A Biological Approach*, 4th edition.

INTRODUCTION

In many parts of the world people do not have access to clean potable (drinkable) water. In more developed countries, water purification systems are a vital part of daily living.

In a cycle known as the **hydrologic cycle**, water is constantly cycled through the process of evaporation, condensation, and precipitation. Precipitation in the form of rainfall or melting snow and sleet eventually winds up (through a process called **runoff**) in storage in either surface water or groundwater. Water collects in surface storage (surface water) in streams that empty into larger streams that empty into ponds, lakes, or oceans. By far the largest storage of surface water is in the oceans. Unfortunately, ocean water contains salts that render it unfit for drinking, irrigation, and most manufacturing needs. The remaining water, called freshwater, sustains the life of most of the earth's plants and animals. **Groundwater** (also known as phreatic water) is water that is under the surface of the ground and makes up the water table that supplies wells. This water is stored in rock, sand, or gravel formations called aquifers. These formations are permeable (allow water to pass through) and are saturated with water. As rainwater falls, it hits the ground and is either absorbed into the ground or begins to run off the surface of the ground. As runoff water goes across the ground, some is absorbed into the ground, and the rest winds up in surface water storage such as lakes and streams. Water pollution occurs in both surface water and in ground water. **Pollution** is the presence of substances in water, air, or soil that impair usefulness or render it offensive to humans.

After completing this exercise, you will have a greater appreciation of water purification systems.

Supplies Needed

- Pond water
- Medicine droppers
- Glass microscope slides and cover slips
- Microscopes with 30× and 100× power
- Internet access or a pond water organism field reference book

Activity Steps

1. Each student will collect a small jar of water from a local pond. Allow the water to sit in the classroom near a window for a few days prior to the lab for incubation. *Note:* River and stream water should not be used; they will not produce favorable results.
2. Place one drop of water onto the slide. Place your cover slip onto the slide by holding the cover slip at a 45° angle to the water and then gently drop it onto the slide.
3. Examine the slide under the microscope at 30×. When you have found an organism, adjust the microscope to 100×.

Using the Internet or reference books, try to identify your organism. To find a good identification key on the Internet, use keywords "pond water key." Guides can also be found under the "Images" browsers using key words "pond water organisms."

Observations

1. Draw a picture of three organisms you found.

2. How does each organism move?

3. Identify any organelles you may be able to see, and label them on your original drawing.

Conclusions

1. What are the names of the organisms you found?

2. Do you think these organisms are single-celled or multicellular? Why?

3. Compare and contrast the organisms you found.

4. Do you think the organisms you found may be harmful to humans?

5. On a scale from 1 to 10, how important are water purification systems? Why?

GLOSSARY OF TERMS

groundwater: Water that is stored under the ground.

hydrologic cycle: The complete cycle through which water passes, commencing as atmospheric water vapor, commencing as atmospheric water vapor, passing into liquid and solid forms as precipitation, into the ground surface, and finally again returning in the form of atmospheric water vapor by means of evaporation and transpiration.

pollution: The presence of substances in water, air, or soil that impair usefulness or render it offensive to humans.

runoff: Water that flows across the ground after a rain.

EXERCISE 19

Wildlife Habitat Suitability: How Does Your Schoolyard Stack Up?

OBJECTIVES

Upon completion of this lab exercise, you should be able to:

- establish a sampling pattern in a wooded area.
- define and measure percent canopy cover using the point-intercept method.
- determine the habitat suitability score for two wildlife species.
- use the habitat suitability data they collect to make recommendations on how to maintain or improve habitat for two wildlife species.

Suggested Reading:

Chapter 19 of *The Science of Agriculture: A Biological Approach*, 4th edition.

INTRODUCTION

Habitat suitability index (HSI) models are tools used by wildlife managers when making decisions about the quality of habitat for different wildlife species. HSI models enable wildlife managers to determine how well suited a habitat is likely to be for a particular animal by providing a numeric score between 0 and 1. Scores closer to 1.00 indicate a greater potential for a particular species to thrive in that habitat. These scores are based on the presence of habitat component, which are essential to the survival of a species. For wildlife species in the southeastern United States, these habitat components could include:

- Percent canopy cover of overstory trees
- Percent canopy cover of overstory trees that produce mast (such as acorns and nuts)
- Average diameter at breast height (DBH) of overstory trees
- Number of trees of a minimum DBH per acre
- Number of snags (standing dead trees) per acre

Percent canopy cover refers to the amount of sunlight that can shine through the branches of overstory (the tallest) trees in the forest. The denseness of the branches of overstory trees has been found to influence the suitability of habitat for various wildlife species, including gray squirrels and barred owls. Follow the steps below to measure the percent canopy cover at different locations around your school.

Supplies Needed

- Paper
- Writing instrument
- Clipboard
- Ruler
- Calculator
- HSI graphs at the end of this activity

Activity Steps

Using the HSI models (graphs) at the end of this activity, determine the suitability of your schoolyard habitat for (a) gray squirrels and (b) barred owls based on the percent canopy cover of overstory trees in a wooded area near your school. Barred owls have been known to prey on gray squirrels, so let's determine the habitat quality of the wooded area near your school for both species.

1. Determine at least three different locations within a wooded area near your school where you are interested in measuring habitat suitability (e.g., in a low-lying area, on top of a hill, and near a stream or river).
2. In each of the locations, get into groups of nine students and form a cross or T on the ground by creating two lines of students that intersect (see figure below).

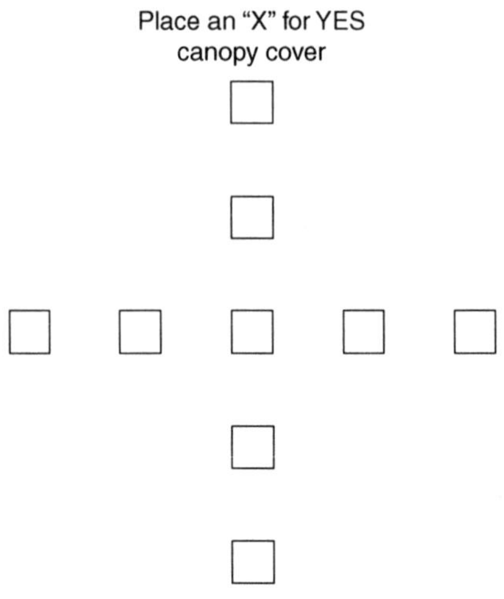

3. Once in "sampling formation," each student will hold a ruler vertically in the air as high as he or she can. Assuming the end of the ruler emits a beam of light, each student will determine if his or her ruler's beam of light would intersect with the leaves in the canopy from where each is standing (without moving). The student will place an X in the boxes on the above figure if the imaginary beam of light intersects with the leaves/branches of the canopy. This is referred to as the **point-intercept method** of sampling.
4. Using your calculator, add up the total number of X's in the boxes and divide by nine (or the total number of students in the sampling formation). This will indicate percent canopy cover in that location. For example, if you recorded seven X's (and two "no's") the percent canopy cover would be 7 X's / 9 total sampling points = 78% canopy cover in that location.
5. Repeat this sampling pattern in each of your selected locations.
6. Back in the classroom: Calculate the average percent canopy cover across your sampling locations.
7. Back in the classroom: Using the HSI models (graphs) at the end of this activity, determine the suitability of the wooded habitat for gray squirrels and barred owls based on the average percent canopy cover you calculated previously.

Questions for Thought

Given your habitat suitability index findings for the gray squirrel and barred owl, write a short (one page) press release for your local newspaper on your findings. Be sure that your press release answers the following questions:

a. Based on your habitat suitability scores, is the wooded area near your school suitable for gray squirrels? Is it suitable for barred owls?
b. How did percent canopy cover change between your different sampling locations? Was one location particularly good for gray squirrels? Was one location particularly good for barred owls? Why?
c. If you were the wildlife manager in charge of this wooded area, what would you recommend to improve or sustain this habitat for gray squirrels and barred owls? Do they need more or less canopy cover?
d. If percent canopy cover was reduced to 25% in this wooded area, would gray squirrel habitat quality be impacted? If so, how?
e. If percent canopy cover was reduced to 25% in this wooded area, would barred owl habitat quality be impacted? If so, how?

Gray squirrel HSI Model

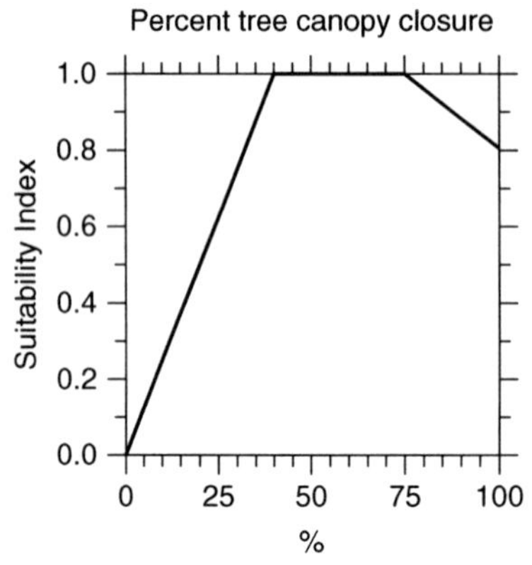

Barred owl HSI Model

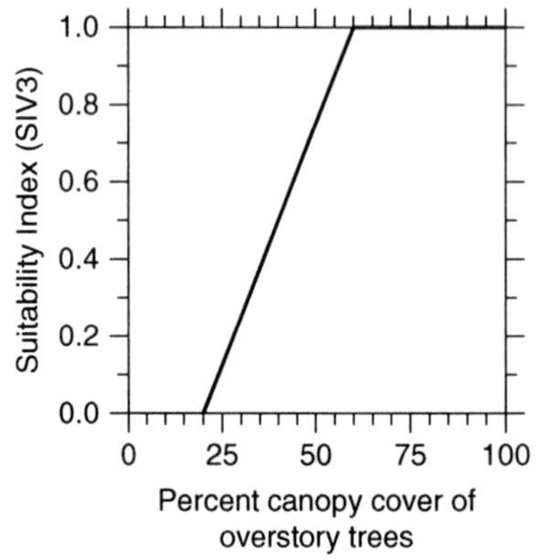

GLOSSARY OF TERMS

Habitat suitability index (HSI) model: A tool used by wildlife managers to determine the quality of habitat for specific wildlife species; provides a habitat quality (suitability) score between 0 and 1.00, with scores closer to 1.00 indicating better habitat quality.

Percent canopy cover: The denseness of branches and leaves of the majority of trees in a wooded area; typically associated with the amount of sunlight able to penetrate through the tree canopy to the forest floor.

Point-intercept method: A technique used by natural resource managers to measure percent canopy cover by determining how often an imaginary point along a line intersects with the forest canopy.

EXERCISE 20

Food Microbiology: Milk and Milk Products

OBJECTIVES

Upon completion of this lab exercise, you should be able to:

- name some of the bacteria commonly present in raw milk and describe their effects on milk quality.
- use the Levowitz-Weber stain reagent to examine milk cultures from a cow with mastitis; identify the causative agent, Streptococcus agalactiae.
- use the Gram stain technique to determine whether mastitis milk bacteria are Gram-positive or Gram-negative.
- determine the antibiotic sensitivity of mastitis milk bacteria to ampicillin, tetracycline, penicillin, and erythromycin and make recommendations for appropriate antibiotic therapy based on the results.
- inoculate China Blue lactose agar plates with yogurt containing active bacterial cultures and identify the lactose-metabolizing bacterial colonies.
- describe the process of heat pasteurization as it applies to raw milk.
- estimate the number of bacterial colonies present in raw milk using the serial dilution plating technique.
- determine the minimum length of time necessary for adequate pasteurization of raw milk using a predetermined dilution factor and the standard plate count method.

Safety Note:

The chemicals used in the Levowitz-Weber stain reagent are volatile and toxic. Be certain that the room is well ventilated or use the staining reagent in a fume hood. Safety goggles with splash guards should be worn all times when handling chemical reagents. Check with your instructor regarding the proper disposal of chemical wastes. When handling raw milk samples, it is always a good idea to wear gloves and to wash your hands thoroughly with antibacterial soap when you are finished.

Suggested Reading:

Chapters 13, 20, and 21 of *The Science of Agriculture: A Biological Approach,* 4th edition.

INTRODUCTION

Prior to the introduction of modern pasteurization techniques and food handling protocols, raw milk was a source for infectious disease bacteria. Milk from any mammal has an endemic microfloral population present at the time of milk collection. The milk can become contaminated with additional bacteria during collection and handling. Soil, hay, and manure in the dairy barn are potential contamination sources. Milk collection equipment, utensils, and dairy workers are also likely sources of microfloral contamination. Public health agencies working in conjunction with the United States Department of Agriculture (USDA) and the Food and Drug Administration (FDA) have set sanitary standards for pasteurized milk and noncultured dairy products in the marketplace (see Table 20-1).

Alteration of the taste and smell of milk and milk products occurs when some microbial population numbers increase to the point that their metabolic by-products are discernible. Milk sours and coagulates when large numbers of lactose-fermenting bacteria are present. *Streptococcus lactis* may account for 90% of the total microflora in freshly soured milk (Harrigan and McCance, 1976). Gas bubbles may impart a frothy appearance to the cream layer or milk surface; the gas bubbles result from the activities of coli-aerogenes bacteria, particularly *Enter-obacter aerogenes*. Lactose-fermenting yeasts may also be present. When heat-treated milk curdles, *bacillus* species are the most likely culprit. When the **viscosity** of refrigerated milk increases to the point where the milk can be drawn out into threads using a wire loop, capsulate bacteria such as *Klebsiella aerogenes, Alcaligenes viscolactis,* and capsule-forming strains of *Bacillus subtilis* and *Micrococcus,* are suspect. When the cream breaks into smaller particles and will not reemulsify in the milk, bacteria-producing lecithinase are present *(Bacillus cereus*

Milk Product	Bacterial Population Limit
Raw milk	200,000 to 400,000/ml using standard plate count 50 to 150/ml using coliform plate count
Grade A pasteurized milk	< 20,000/ml using standard plate count < 5/ml using coliform plate count

Table 20-1. Standards for grade A raw and pasteurized milk-bacterial count limits.

and *Bacillus cereus,* var. *mycoides).* A malty taste can be imparted to milk by the activities of *Streptococcus lactis,* var. *maltigenes.* Phenolic taints arise from the action of *Bacillus circulans.* The endospores of *B. circulans* can survive heat treatment, and the carbolic or phenolic taste can be observed in pasteurized milk.

EXAMINATION OF MASTITIS MILK

Mastitis is the inflammation of the udder of a female domesticated animal. All female mammals are susceptible, but mastitis is most frequent among cows, goats, and sheep. The inflammation may result from an infection caused by *Streptococcus agalactiae. S. agalactiae* is one cause of infant bacterial meningitis in humans, but in livestock, the infection is usually restricted to the udder. You will use a stain, Levowitz-Weber stain reagent (DiLiello, 1982) to examine raw milk from a cow with mastitis. You will need to follow the directions carefully in order to obtain a good milk smear and stain. Wear safety goggles and gloves when handling the milk and stain reagent.

Preparation of Bacterial Smears for Staining

1. Holding a disposable sterile inoculating loop like a pen, dip the loop into the mastitis milk culture provided by your instructor.
2. Transfer the drop of milk to the surface of a clean, flat slide.
3. Using the flat side of the loop, spread the milk droplet over the entire surface of the slide. The goal is to obtain a *thin, evenly spread smear* that is almost invisible to the unaided eye.
4. Let the slide air dry completely, 3 to 10 minutes depending on the viscosity of the milk sample.
5. Using a wooden clothespin to hold the slide, pass the slide through the flame of a Bunsen burner or alcohol lamp twice slowly to heat-fix the bacterial cells to the surface of the slide.
6. Allow the slides to cool completely before staining.
7. Repeat the process until you have *four* mastitis milk smears, two for use with the Levowitz-Weber stain reagent and two for the Gram staining procedure. Always use a fresh sterile inoculating loop when preparing the next slide.

Remember to always handle your slide with a clothespin during the staining procedure until the slide is completely dry.

NOTE:
You will need to use a microscope with an oil-immersion lens to make your observations. For a review of the proper use of an oil-immersion lens, see Exercise 3 of this manual.

Levowitz-Weber Staining Reagent

1. In a well-ventilated area, fill a Wheaton staining jar (either vertical or horizontal) to the top so that any slides placed in the jar will be completely submerged. *Always* make certain that the glass lid is on the jar when you are not adding or removing slides from the staining jar.
2. Remove the lid from the staining jar and completely immerse the mastitis milk smear slides in the Levowitz-Weber dye solution by slipping the ends of the slide between the grooves on the inside of the staining jar. Replace the lid.
3. Allow the slides to remain immersed in the dye for 2 minutes. Do not leave the slides for a longer period of time; this results in overstaining!
4. After 2 minutes, remove the staining jar lid, and pull the slides from the solution using a wooden clothespin. Replace the lid on the staining jar.
5. Place the slide in a slide holder oriented so that the excess stain drains away from the slide and allow it to air dry.
6. When your slide has completely dried, gently rinse your stained smear slide in three separate rinse baths of warm tap water (35°C–45°C).
7. Allow the slides to air dry completely before viewing them under the microscope. Do not blot your milk smear slides with paper or tissue as you may remove cells or reduce the number of cells on the slide.

After initially focusing your slide with the coarse adjustment knob, use the fine adjustment knob to make any alterations in the focus of the slide. Bacterial clumps, white blood cells, and body cells (from the lining of the cow's udder) will be dark blue in color. Milk protein will appear light blue in contrast to the unstained, colorless fat globules. Any dirt or dust will be dark brown to black in color. Remember from Exercise 6 that the prefix *strepto* refers to bacteria that form large chains of cells. **Coccus** are spherical bacterial cells. Round cells in long chains characterize members of the genus *Streptococcus*. Draw and label what you observe under 1,000× magnification in the space below. Label bacterial cells (*Streptococcus* if you see any), fat globules, and milk protein. See Figure 20-1 for a visual reference.

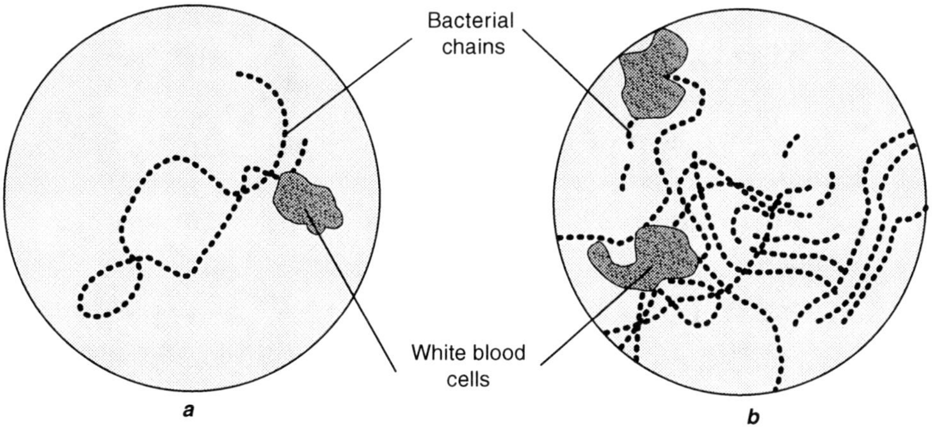

Figure 20-1. (a) Visual inspection of mastitis milk from a cow. (a) Moderate case of mastitis milk. (b) Serious case of mastitis milk. Both are mastitis milk at 1000×. If source of the mastitis symptoms is Streptococcus agalactiae, long chains of blue spherical cells are visible. Larger, irregularly shaped white blood cells will also be common. Good quality milk (not shown) will have white circular areas visible in the light blue background. These are fat droplets which are insoluble in the stain. A few white blood cells or bacteria may be visible but not abundant.

Gram Stain

The Gram stain is the single most important stain used in bacterial identification, dividing bacteria into two major groups, Gram-positive and Gram-negative. A Danish physician, Hans Gram, developed the staining technique. He noticed that certain bacteria, after being stained with crystal violet, become decolorized when washed with ethyl alcohol. These bacteria are referred to as **Gram-negative**, while those that retain the violet stain under these conditions are called **Gram-positive**. The primary stain, crystal violet, is followed by an iodine solution that helps fix the primary dye to Gram-positive organisms. The smear is then decolorized with 95% alcohol and counterstained with a red dye such as safranin, which stains Gram-negative organisms.

Gram Stain Procedure

1. Use the two remaining mastitis milk smear slides you prepared earlier for this portion of the exercise.
2. Fill a clean Wheaton staining jar with crystal violet stain. Use a clothespin to manipulate the slide, completely submerging the slide in the crystal violet stain. Place the lid on the staining jar.
3. Allow the mastitis milk smear slide to stand in the crystal violet solution for 1 to 2 minutes—*no longer*.

4. Remove the lid from the staining jar. Using the clothespin, remove the stained smear slide from the crystal violet and replace the jar lid.
5. Rinse the slide rapidly with warm water.
6. Fill a clean Wheaton staining jar with Gram's iodine solution. Use a clothespin to manipulate the slide, completely submerging the slide in the iodine solution. Place the lid on the staining jar.
7. Leave the slide in Gram's iodine for 1 minute—*no longer.*
8. Remove the lid from the staining jar, using the clothespin, remove the stained smear slide from the iodine solution and replace the jar lid.
9. Place the slide in a slide holder oriented so that the excess stain drains away and blot dry
10. Wash the slide with 95% ethanol until no more purple stain washes off the slide (5 to 20 seconds for most thin smears).
11. Rinse the slide with warm water.
12. Fill a clean Wheaton staining jar with carbol fuschin staining solution. Use a clothespin to manipulate the slide, completely submerging the slide in the carbol fuschin stain. Place the lid on the staining jar and allow the slide to stain for 20 seconds—*no longer.*
13. Rinse the slide well and blot dry.

Bring the slide into focus first using the low-power objective lens and then the high-power objective lens. Place a drop of immersion oil on the slide and focus using the oil immersion lens. Remember that you do not want to get immersion oil on the high-power objective lens, and immersion oil should always be removed from the oil-immersion objective lens with a cleaning solution when you are finished.

Are the bacterial cells in the mastitis milk Gram-positive or Gram-negative? How do you know? (*Hint:* Look at the color.)

What is the principal cell wall component of the bacteria you see? (See Table 6-1 in Exercise 6).

ANTIBIOTIC SENSITIVITY OF MASTITIS MILK BACTERIA

Many cases of mastitis in cows, goats, and ewes are attributable to bacterial infections requiring the administration of **antibiotics** to eliminate the causative organism(s) from the udder. Veterinarians frequently utilize antibiotic sensitivity tests to determine the relative effectiveness of assorted antibiotics against bacterial pathogens. You will inoculate petri plates containing growth medium with 1 ml of mastitis milk using a sterile technique. Four different antibiotic disks, ampicillin, erythromycin, penicillin, and tetracycline will be applied to the surface of the inoculated petri plate to evaluate the efficacy of each drug in treating mastitis.

Hypothesis

Write a hypothesis that relates the relative sensitivity of *Streptococcus agalactiae* to ampicillin, erythromycin, penicillium, and streptomycin based on bacterial cell wall composition. (*Hint:* Look at the results of the Gram stain procedure and Table 6-1 in Exercise 6.)

Safety Note:

The characteristics of microbial organisms, small size, speedy development, and rapid reproduction are useful to the student who is using these organisms as a study tool. These are extremely advantageous to microbes and explain the biological success of bacteria and fungi. These microbial attributes can cause serious infection to develop. Therefore, you must exercise care and precision when handling microbes. *You must follow EVERY step in the inoculation or handling procedure in order.*

Important Safety Rules

1. Food must not be eaten in the laboratory.
2. Keep all objects such as fingers, paper, and pencils out of your mouth.
3. Never drink from laboratory glassware.
4. Use **aseptic technique (sterile)** whenever you work with bacteria.
5. If a living culture is spilled, notify the instructor immediately and disinfect your hands.
6. Discard all cultures *only* according to the instructions of your lab instructor.

7. Liquid waste and agar should be disposed of only in the plastic bags provided for this purpose.
8. At the conclusion of the work period, return all equipment to its place. Wash the desk tops and your hands with disinfectant.

Aseptic Technique

You must maintain sterile conditions when transferring and culturing microorganisms. Read these instructions several times before attempting to transfer bacteria.

Follow these instructions very carefully:

Obtain a petri dish containing special plate count agar media (Oxoid Unipath).

1. Turn the petri dish over so that the agar side is on the top. Using a permanent marker, mark the outside bottom of the dish, dividing the area into four each "pie pieces" or quadrants. Label one quadrant *amp* (for ampicillin), one quadrant *eryth* (for erythromycin), one quadrant *pen* (for penicillin), and the last quadrant *tetra* (for tetracycline). Write your name and the date on the bottom of the dish, and return the petri dish to the upright orientation (agarose side down).
2. Remove a sterile swab from its wrapper.
3. Open a sterile test tube containing mastitis milk and flame the mouth of the tube.
4. Insert the swab into the milk; squeeze out excess fluid.
5. Remove swab, flame the mouth of the test tube again, and return the test tube to the rack.
6. Open an agar petri dish lid *very slightly;* allow space *only* for swab to pass.
7. Smear the entire agar plate with the swab. Do half the plate, then turn a quarter of a turn and repeat three times (see Figure 20-2).
8. Close the petri dish and discard the swab into disinfectant.
9. Pass the tips of a pair of forceps through the flame of an alcohol lamp or Bunsen burner to sterilize the tips—DO NOT lay the sterile forceps down on the bench/table. Open an agar petri dish lid very slightly; allow space only for forceps to pass. Carefully remove an ampicillin antibiotic disk from the container and place it on the surface of the agar in the quadrant labeled *amp*. Using the forceps, gently press the antibiotic disc into the agar. Close the petri dish, and flame the tips of the forceps.
10. Repeat this process until all four antibiotic disks are on the agar surface in the appropriately labeled quadrants.
11. Incubate the plates agar side up at 37°C for 48 hours in an incubator or at room temperature for four days in a drawer.

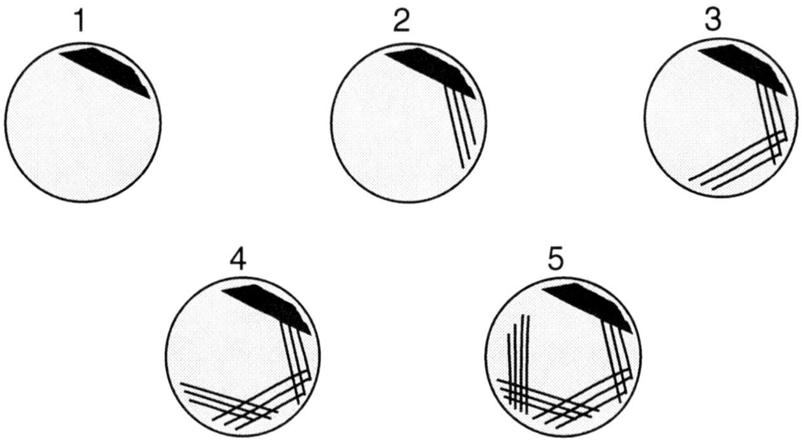

Figure 20-2. Technique for streaking agar petri dishes with micorbial inoculant.

Antibiotic	Zone of Inhibition (mm)
Ampicillin (amp)	
Erythromycin (eryth)	
Penicillin (pen)	
Tetracydine (tetra)	

Table 20-2. Results of antibiotic sensitivity tests for four different drugs on mastitis milk bacteria.

12. At the end of the incubation period, check the plates for inhibition of bacterial growth. If the antibiotic is effective against the mastitis bacteria, a halo (or **zone of inhibition**) will encircle the disk. This clear area represents a zone of no growth—the size the zone represents the strength of inhibition.
13. Using a metric ruler, measure the zones of inhibition around each of the four antibiotic disks and record the results in Table 20-2.

Results

Which antibiotic was most effective in inhibiting microbial growth?

Did the results support or disprove your hypothesis?

You are a veterinarian asked to examine and treat several cases of mastitis in a dairy barn. Describe how you would use direct observation (Levowitz-Weber and Gram-staining techniques) and antibiotic sensitivity testing in the diagnosis and treatment of mastitis.

BENEFICIAL MICROORGANISMS IN THE DAIRY INDUSTRY

Some bacteria are purposefully added to dairy products to produce a desired taste, texture, or product. Cheeses are aged as a result of bacterial and fungal activities. Buttermilk, sour cream, and yogurt derive their characteristic flavor and texture from bacterial action. Some people experience difficulties in digesting milk and dairy products due to a condition called lactose intolerance. Lactose-intolerant individuals lack the digestive enzyme, lactase, necessary for the metabolism of milk sugar (lactose). Consumption of lactose-containing products can produce bloating, pain, and diarrhea in lactose-intolerant persons. Yogurt and acidophilus milk contain bacteria that belong to the genus *Lactobacillus*. *Lactobacillus* utilizes lactose as an energy source for growth, and products containing *Lactobacillus* are well tolerated by lactase deficient people. Table 20-3 summarizes the uses of some bacteria in the commercial production of dairy products.

In this experiment, you will employ sterile techniques to inoculate petri dishes containing China Blue lactose agar with active cultures from a commercially available brand of yogurt (such as Dannon, plain). If *Lactobacillus* is present, the bacteria will incorporate the blue dye into the bacterial cell during the process of utilizing the lactose in the media. *Lactobacillus* colonies will be blue in appearance. Nonlactose fermenting bacterial colonies will be white in appearance.

Culture	Food Use	Function
Lactobacillus acidophilus and *Streptococcus thermophilus*	Acidophilus milk, yogurt, Swiss and Emmenthaler cheese	Acid
Lactobacillus bulgaricus, L lactis, and *L helveticus*	Yogurt, kefir, kumiss, Swiss and related cheeses	Acid and flavor
Leuconostoc cremoris	Cultured buttermilk, butter, and cottage cheese	Flavor
Propionibacterium freundenreichu	Emmenthaler, Swiss, and related cheeses	Flavor and eye formation
Streptococcus cremoris	Cottage cheese and cultured buttermilk	Acid and flavor
Streptococcus lactis	Cottage cheese and cultured buttermilk	Acid and flavor
Streptococcus lactis subspecies *diacetyllactis*	Butter, sour cream, and cultured buttermilk	Acid and flavor

Table 20-3. Major bacterial species used in the commercial production of dairy products.

Procedure

Obtain a petri dish containing China Blue lactose agar media.

1. Turn the petri dish over so that the agar side is on top. Using a permanent marker, label the outside bottom of the dish with your name, the date, and yogurt culture on the bottom of the dish. Return the petri dish to the upright orientation (agarose side down).
2. Remove a sterile swab from its wrapper.
3. Open a sterile test tube containing the yogurt cultures and flame the mouth of the tube.
4. Insert the swab into the yogurt; squeeze out excess fluid.
5. Remove swab, flame the mouth of the test tube again, and return the test tube to the rack.
6. Open an agar petri dish lid *very slightly;* allow space *only* for a swab to pass.
7. Smear the entire agar plate with the swab. Do half the plate, then turn a quarter of a turn and repeat three times (see Figure 20-2).
8. Close the petri dish and discard the swab into disinfectant.

9. Incubate the plates agar side up at 37°C for 48 hours in an incubator or at room temperature in a drawer for four days.
10. Inspect the Chine Blue lactose plates at the end of the incubation period for blue and white colonies. Count the number of blue colonies and white colonies and record the number in the space provided.

Results

Compare the color of the incubated plates to the original color of the plates prior to inoculation. Do you see any differences?

What does this indicate about the number of lactose fermenting bacteria in yogurt?

PASTEURIZATION

Pasteurization is a specific type of food preservation technique involving heating food to a temperature less than 100°C (the temperature at which water boils) for an extended period of time. Pasteurization is used in the manufacture of beer, wine, vinegar, baked goods, cured meats, and milk products. The hot temperatures used in the pasteurization process kill the vegetative bacterial cells, but the heat-resistant bacterial endospores of some pathogenic organisms survive. Modern pasteurization techniques have greatly reduced the frequency of food-related illnesses.

There are two different milk pasteurization processes utilized in modern dairies. Early bulk operations used low-temperature holding (LTH) to pasteurize large quantities of milk in a vat. The milk was heated to 62.9°C for 30 minutes to ensure that any vegetative cells of *Mycobacterium tuberculosis* or *Coxiella burnetti* are destroyed. The high-temperature short-time (HTST) pasteurization procedure is more compatible with modern, mechanized diaries. During HTST processing, the milk is heated for 15 seconds at 71.5°C guaranteeing adequate control of pathogens (see Figure 20-3).

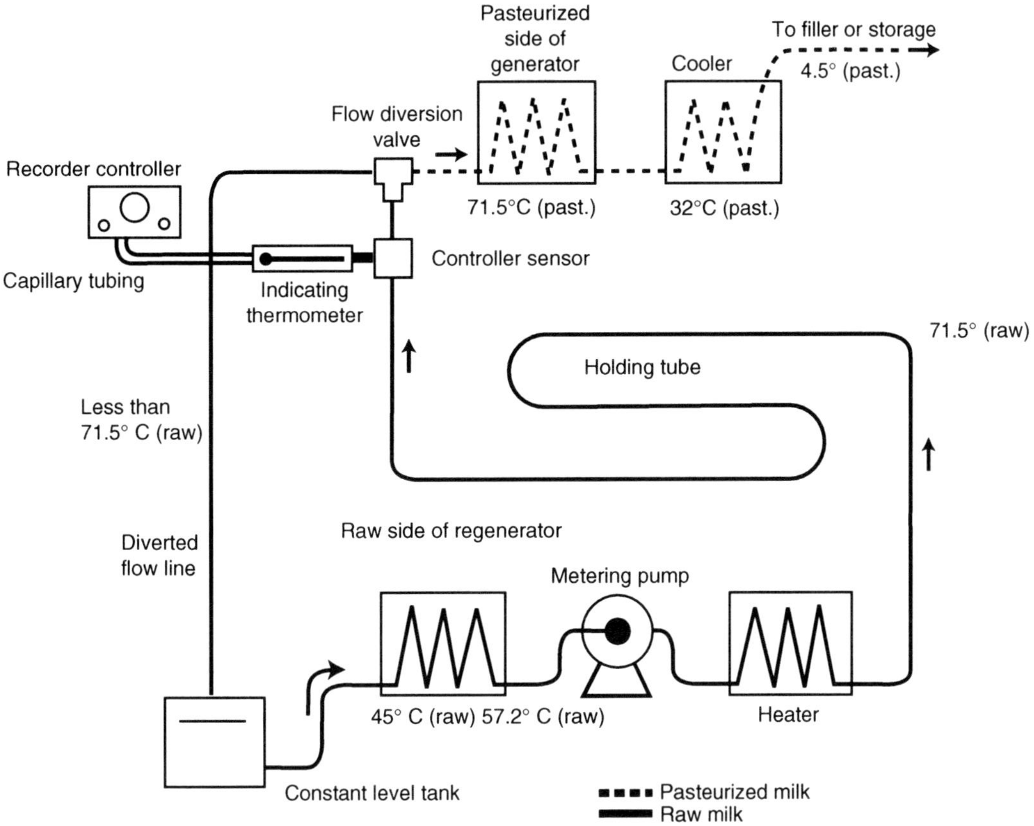

Figure 20-3. High-temperature short-time processing for milk pasteurization.

ESTIMATION OF BACTERIAL COLONY NUMBER IN RAW MILK

Bacterial colony counts are used to grade raw milk. Grade A raw milk should not exceed 100,000 colonies per plate (1 g sample). In contrast, Grade 1 milk, which is used in the making of dairy products such as cheeses, butter, and yogurt, may have a plate count as high as 500,000 colonies per plate (1 g sample). You are going to use the serial dilution plating technique to estimate the number of bacterial colonies present in 1 ml of raw milk or buttermilk with active cultures.

1. Remove the stopper from a sterile test tube and add 1 ml of milk to 10 ml of sterile distilled water. Remember that you do not want to touch the bottom of the stopper or lay it on your table or lab bench; you want to avoid potential contamination of your milk-sterile water solution. Put the stopper back in place and shake the test tube so that the milk is suspended throughout the water column.
2. Allow the large particles to settle to the bottom of the test tube.

3. Obtain three test tubes, each containing 9 ml of sterile water and label them: 1/10, 1/100, and 1/1,000. Make a serial dilution of your original milk-sterile water sample as shown in Figure 5-4 (Exercise 5) in this manual. Use the same pipette for all three transfers, but rinse it between each transfer. In other words, after discharging the contents of the pipette into the new dilution tube, rinse the pipette by drawing up some of the milk/water and emptying the pipette again. Remember to mix each tube before continuing the dilution.
4. Label six sterile petri dishes (two each) 1/10, 1/100, and 1/1,000 and add *your initials, date,* and *raw milk.*
5. Using a clean pipette and beginning with the 1/1,000 dilution, inoculate each of the six plates with 1 ml from the appropriate dilution tube.
6. Obtain a flask of molten special plate count agar from the heated water bath or a microwave. Using aseptic technique, pour enough agar into one of the inoculated dishes to cover 2/3 of the plate. Immediately mix the liquified agar with the diluted water sample by gently rotating the plate clockwise and counterclockwise without splashing or lifting it from the lab bench. Repeat the process with the other inoculated petri dishes. Allow the plates to cool until the agar is well hardened. Place a piece of tape on either side of each dish to prevent them from coming open. Incubate the plates at 37°C for 48 hours or 4 days at room temperature.
7. After the plates have had sufficient time to incubate, count the colonies present on those plates where counting is feasible.

In the space below, record the number of colonies present on each plate and calculate the number of viable bacteria present per ml of the original milk-water sample, taking into account the predilution performed on the 1 ml milk sample (10 ml of sterile water). (Students often wonder why equations are not provided to complete the calculations. The answer is that by reasoning it out for yourself and coming up with your own formula, you will have a much better understanding of what is involved than you would if you simply plugged your data into a formula.)

Predilution, if any, for your sample (A, B, & C) *1 ml raw milk/10 ml sterile water.*

Colonies present per plate:
1/10 _____ 1/100 _____ 1/1,000 _____

Viable bacteria present per ml of the *original* raw milk sample: _____ colonies/ml

Which dilution factor gave the best colony count (the dilution which produced visible individual colonies evenly distributed on the plate)?

Pasteurization Time

Milk that is properly pasteurized should have no more than 20,000 colonies per gram in order to receive the designation Grade A. The length of time that the milk is treated is critical to the reduction of bacterial numbers. In this segment of the lab exercise, diluted raw milk samples will be heated to 63°C for specified lengths of time to determine the minimum time necessary to ensure adequate protection from bacterial contaminants. The standard plate count method will be employed for the estimation of bacterial numbers.

1. Remove the stopper from a sterile test tube and add 1 ml of raw milk to 10 ml of sterile distilled water. Remember that you do not want to touch the bottom of the stopper or lay it on your table or lab bench; you want to avoid potential contamination of your milk-sterile water solution. Put the stopper back in place and shake the test tube so that the milk is suspended throughout the water column.
2. Allow the large particles to settle to the bottom of the test tube.
3. Dilute the milk to the concentration that produced the clearest spread of colonies in the serial dilution procedure above. The milk is currently at a 1/10 dilution rate. If the 1/100 dilution produced the best results, dilute the 1/10 sample by pipetting 1 ml of the 1/10 milk-sterile water solution into 9 ml of sterile water. If the 1/1,000 dilution produced the best results prepare a 1/100 dilution and then a 1/1,000 dilution. Use the same pipette for all three transfers, but rinse it between each transfer. In other words, after discharging the contents of the pipette into the new dilution tube, rinse the pipette by drawing up some of the milk/water and emptying the pipette again. Remember to mix each tube before continuing the dilution.
4. Label eight sterile petri dishes to indicate thermal treatment time intervals: 5 minutes, 10 minutes, 15 minutes, 20 minutes, 25 minutes, 30 minutes, 35 minutes, and 40 minutes. *Include your initials, date,* and *raw milk.*
5. Obtain a flask of molten special plate agar from the heated water bath or a microwave. Using aseptic technique, pour enough agar into one of the inoculated dishes to cover two-thirds of the plate. Immediately mix the liquefied agar with 1 ml of milk by gently rotating the plate clockwise and counterclockwise without splashing or lifting it from the lab bench. Label the plate 0 minutes.
6. Pipette 20 ml of raw milk into a sterile test tube. Label your test tube with your initials.
7. Place the test tube in a 63°C water bath or beaker and note the time that the heat treatment began. At 5 minutes, use a sterile disposable 1 ml pipette to withdraw 1 ml of heat-treated raw milk.

Obtain a flask of molten special plate count agar from the heated water bath or a microwave. Using aseptic technique, pour enough agar into one of the inoculated dishes to cover 2/3 of the plate. Immediately mix the liquified agar with 1 ml of milk by gently rotating the plate clockwise and counterclockwise without splashing or lifting it from the lab bench. Label the plate 5 minutes.

8. Repeat the process at 5 minute intervals: 10 minutes, 15 minutes, 20 minutes, 25 minutes, 30 minutes, 35 minutes, and 40 minutes. Use a new sterile disposable pipette to withdraw milk from the heat-treated test tube each time.
9. Allow the plates to cool until the agar is well hardened. Place a piece of tape on either side of each dish to prevent them from coming open. Incubate the plates at 37°C for 48 hours or 4 days at room temperature.
10. After the plates have had sufficient time to incubate, count the colonies present on those plates where counting is feasible and record your results in Table 20-4. Remember to multiply your plate counts by the dilution factor to estimate the total number of colonies present in 1 ml of milk.

Time (minutes) at 63°C	Estimated Bacterial Colonies/ml Raw Milk 1
Dilution Factor: ______	
5 minutes	
10 minutes	
15 minutes	
20 minutes	
25 minutes	
30 minutes	
35 minutes	
40 minutes	

Table 20-4. Bacterial plate counts for diluted raw milk as a function of time.

Results

Construct a line graph with the treatment time (minutes) on the horizontal (x-axis) and the number of bacterial colonies on the vertical axis (y-axis). At what time point (in minutes) does the number of colonies drop below 20,000 (level for Grade A milk)?

What is the minimum time necessary for good control of bacterial contaminants by thermal treatment at 63°C?

Why would it be important to determine the *shortest* length of time necessary for adequate control of bacteria in milk and milk products? (*Hint:* Think about the factors that affect profitability and product quality.)

GLOSSARY OF TERMS

antibiotic: (*anti*—against; *biotikos*—pertaining to life) An organic molecule produced by a microorganism that inhibits or retards the growth of other microorganisms; may be produced synthetically from derivatives of microorganisms.

aseptic technique: (*a*—without; *septikos*—putrefying) Laboratory method for working with microorganisms involving meticulously clean technique; hands must be washed and gloved, lab bench surfaces are cleaned prior to and after working with any living cultures, culture loops are flamed before and after working with cultures, and the mouths of culture tubes are flamed before the tube is inoculated and following inoculation; also referred to as sterile technique.

bacillus: (pl. *bacilli*) A rod-shaped bacterium.

coccus: (pl. *cocci*) A spherical-shaped bacterium.

Gram-negative: Bacterial cells which appear pink when the Gram-staining technique is used.

Gram-positive: Bacterial cells which stain purple when the Gram-staining technique is used.

mastitis: (*itis*— inflammation of) An inflammation of the udder of a female domestic animal; may be infectious (*Streptococcus agalactiae)* or noninfectious in origin; results in decreased milk production, fever, swelling and tenderness of the udder; most common among cows, goats, and ewes.

microflora: (*micro*—small; *flora*—plants) Microscopic plant like organisms, including bacteria, fungi, and protists; native to an area.

pasteurization: Any thermal food-preserving process in which the highest temperature reached is below 100°C.

prokaryotic: (*pro*—first; *karyon*—kernel) A cell lacking a membrane-bound nucleus or membrane-bound organelles; bacteria and cyanobacteria have prokaryotic cells.

spirillum: (pl. *spirilla*) A corkscrew or spiral-shaped bacterium.

viscosity: The state of having a glutinous or gelatinous consistency; sticky.

zone of inhibition: Distance around an antibiotic (or other antimicrobial substance) disc placed in a microbial culture in which no microbial growth is observed.

EXERCISE 21

Energy from Agriculture: Making Biodiesel

OBJECTIVES

Upon completion of this lab exercise, you should be able to:

- identify three ways to use vegetable oil to run a diesel engine.
- list three benefits of biodiesel.
- describe the different physical characteristics of vegetable oil versus biodiesel.

Suggested Reading:

Chapter 25 of *The Science of Agriculture: A Biological Approach*, 4th edition.

INTRODUCTION

Biodiesel is an alternative fuel source that many consider to be one of the most promising new energy sources. It has already been widely used across the world for some time. One benefit to using biodiesel is fewer harmful emissions when compared to cars running on petroleum-based fuels. Another is that vegetable oil is a renewable fuel source. Biodiesel is made from fats or lipids that come from vegetable oils or animal fats and combining them with alcohol.

Today in the United States the most commonly used sources for biodiesel are soybean oil and recycled frying oil from the restaurant industry. Biodiesel has become popular over the past several years because of the ease of making diesel engines compatible with biodiesel. There are multiple ways to run a diesel engine on vegetable oil (each with their own positive and negative aspects):

- Mix it with petroleum diesel fuel at different amounts.
- Use **straight vegetable oil (SVO)** or **waste vegetable oil (WVO)**.
- Convert SVO or WVO to biodiesel.

Mixing vegetable oil with petroleum-based fuels at different amounts provides a way to make petroleum fuel go farther. Running a car on SVO can be easily done but requires modifications to run properly.

People who drive these converted cars report that their vehicles run just as smoothly as they did before the conversion with the same if not improved fuel mileage. There are both positive and negative aspects to using SVO. In addition, people report that their automobiles give off a slight odor of whatever food was last cooked in the oil. Although it requires additional processing, unlike SVO, biodiesel can be used in any diesel without modification.

For this exercise we will be converting SVO to biodiesel.

Supplies Needed

- 1 liter *new* vegetable oil
- 4 grams substance A (sodium hydroxide or caustic lye)
- 250 milliliters substance B (methanol or wood alcohol)
- 2 glass containers with lids
- 1 funnel
- 4 quart pot
- Warming plate
- Thermometer
- Blender (do not use with food again)
- Turkey baster (do not use with food again)
- Protective eyewear
- Thick rubber gloves

Activity Steps

Safety Note

Do not attempt to make biodiesel at home! Wear safety gloves and eye protection glasses during this exercise, and make sure that all items of equipment are clean and dry.

1. On the warming plate, heat the oil to 55°C in the 4 quart pot and remove it from the heat.
2. Pour the 250 ml of substance B into the glass container and quickly replace the lid. Be careful not to breathe in any fumes.
3. Measure out 4 grams of substance A. Be sure not to let this substance come into contact with any moisture. Using a clean and dry funnel, add substance A to substance B in the glass container. Replace the lid.
4. Keeping the container away from your face, swirl the mixture gently for 1 minute. Check to see if the granules are dissolved. If they are not, swirl again for 1 minute. Repeat this process until all of substance A has completely dissolved (this can take up to six times). The mixture will become very warm as the chemicals combine.

5. Check the blender to make sure that all of the gaskets are in good condition and all parts are tightly fitted. Pour oil into the blender.
6. With the blender turned OFF, slowly add the A/B solution. Secure the blender lid and turn the blender on to a low speed. Allow the solution to mix for 20 minutes.
7. Transfer the solution to a clean glass jar. Within 20 minutes, you will begin to see a separation. After 24 to 36 hours, there will be two distinct layers (biodiesel) and a thick brown layer (glycerin by-product).

Observations

1. Describe the appearance of the three substances prior to mixing.
2. Why do you think mixing substances A and B together produced heat?
3. Describe the appearance of the mixture soon after it was taken out of the blender.
4. Draw a picture of the mixture 20 minutes after it was take from the blender. Label your picture.

Conclusions

1. Did the mixture produce two distinct layers after it had time to settle? If not, what do you think happened?

2. List three benefits of producing biodiesel.

3. As a farmer, what crop(s) could you produce for use in the biodiesel industry?

GLOSSARY OF TERMS

biodiesel: Alternative fuel source made from fats that come from vegetable oils or animal fats and are combined with alcohol.

renewable resources: Resources such as trees that can be replaced.

straight vegetable oil (SVO): Unused vegetable oil used as fuel.

waste vegetable oil (WVO): Vegetable oil used as fuel that has been discarded by the restaurant industry